Parasites, Worms, *and the* Human Body *in* Religion *and* Culture

PETER LANG
New York • Washington, D.C./Baltimore • Bern
Frankfurt • Berlin • Brussels • Vienna • Oxford

Parasites, Worms, *and the* Human Body *in* Religion *and* Culture

EDITED BY Brenda Gardenour
AND Misha Tadd

PETER LANG
New York • Washington, D.C./Baltimore • Bern
Frankfurt • Berlin • Brussels • Vienna • Oxford

Library of Congress Cataloging-in-Publication Data

Parasites, worms, and the human body in religion
and culture / edited by Brenda Gardenour, Misha Tadd.
p. cm.
Includes bibliographical references.
1. Parasitic diseases—Social aspects. 2. Parasites—Psychological aspects.
3. Worms—Psychological aspects. 4. Human body—Religious aspects.
5. Religion and culture. 6. Anthropology—Comparative method.
I. Gardenour, Brenda S. II. Tadd, Misha.
RC119.P347 616.9'6—dc23 2011043817
ISBN 978-1-4331-1547-9 (hardcover)
ISBN 978-1-4539-0263-9 (e-book)

Bibliographic information published by **Die Deutsche Nationalbibliothek**.
Die Deutsche Nationalbibliothek lists this publication in the "Deutsche
Nationalbibliografie"; detailed bibliographic data is available
on the Internet at http://dnb.d-nb.de/.

Cover image reproduced courtesy of Wellcome Library, London

29 Broadway, 18th floor, New York, NY 10006
www.peterlang.com

Acknowledgements

The editors of this volume would like to thank the American Academy of Religion and the Boston University Department of Religion, without whose support the 2008 conference on the Wormy Corpus and this volume would not be possible. We would also like to thank Shigehisa Kuriyama of Harvard for his inspiration, Linda Barnes for her unwavering guidance, and the Wellcome Library for the History of Medicine in London.

CONTENTS

The Power of Parasites and Worms

Misha Tadd

At first glance, worms and parasites may seem an odd subject for a collection of essays on religion and culture. These insignificant beings cannot have had much sway on the grand edifice of human civilization: why then produce such a volume? Why honor the creepy and the crawly with an interdisciplinary body of analyses?

To date, there has not been a concerted effort to bring these creatures out of isolated study in discrete fields like parasitology and literature. Scientific research is rarely juxtaposed to its humanities counterpart. The former tends to disregard cultural variation and the latter can be overly emphatic about difference and historical particularity. This work engages the topic in a broad comparative effort that seeks to elude the twin perils of glossing over divergence and narrowly focusing on peculiarities.

Parasites and worms are perfectly suited to comparative inquiry for two reasons. First, carried by humans across time and space, they have achieved near-omnipresence and have left marks on every community. Inevitably, the biogenic threat they pose has affected, in however great or small a way, the development of those communities. Second, their symbolic power derives from a nearly universal evocation of fear and disgust. Cognizant of their curious figurative potency, J. Z. Smith once identified them as the exemplary "other," an extreme counterpoint to what we deem familiar and good, and concluded that a discussion of this other is inevitably a discussion of the self.[1] Symbolic use of parasites as "the other" appears cross-culturally as well, but each incarnation speaks to a society's own social prejudices and insecurities.

As Dr. Alfred C. Reed said, in the study of parasitism "is written the foundation of human history."[2] This quotation refers to the social parasitism of living off others, but it also brings to mind the intertwining nature of our own past and that of the creatures living within us.

This volume showcases the fascinating interplay between a common experience and its web of related cultural and religious forms. As a collection, it provides an unusual perspective, perhaps a worm's-eye-view, of these myriad symbolic and living systems.

Definitions

This volume relies on the limited English words "parasite" and "worm" to represent phenomena found in many cultures and identified by a range of language-specific terms. While acknowledging the particularities of these various terms, this work employs these two glosses for fruitful comparative discussion. "Parasite" denotes a type of relationship while "worm" refers to the creatures that function as parasites or are merely the lowest forms of life.

The history of these two English expressions reveals notable themes for the broader conversation. "Parasite" comes from an ancient Greek word whose meaning broaches the full range of issues addressed in these essays. Originally, it referred to a religious relationship, but over time came to invoke a social one, a biological one, and finally a medical one. This semantic drift affirms that even the etymology of the term bridges the same categories that comprise this present study of parasites. *Parasitos* (παράσιτος) initially meant a priest's assistant who was fed at the public's expense. Since they received "on the side" (*para*) their "food/grain" (*sito*), the term developed a negative connotation and came to imply someone who would accept humiliation in exchange for food. It seems that almost from the beginning, this term, with its allusions to subordination and sycophantism, served as an insult, a tendency still preserved in contemporary English.

"Parasite" was not applied to worms and other non-human creatures until the eighteenth and nineteenth centuries, when it was no longer a religious or social term and began to designate a kind of biological connection. First used to describe plants growing on other plants, it gradually included animals living off the nutrients of other animals. A parasite is not, then, a category of "thing," but of "relationship." Fittingly, biology classifies creatures as parasites when their method of survival requires such a dependency. They are the living "other" that siphons off and harms the host by taking more than they give. Those most often hosted by humans include bacteria, viruses, molds, fungi, and worms.

All of these parasites are also technically symbiotes, a biological term referring to organisms whose existence is tightly intertwined with that of another. This is a relationship based on an exchange that can either be imbalanced (parasitic) or balanced (mutualistic). Ideally, symbiote should not be conflated with the opposite of "parasite," as sometimes happens in popular discourse. Parasites are symbiotes, too.[3]

Still, "parasite" is sometimes an overarching gloss for parasitic worms which, depending on the cultural context, can be understood as parasitic, mutualistic, or a mixture of both. One can parse these categories even more finely, delving into parasites that live off their hosts (biotrophs) and those that sterilize, kill, or consume their hosts (necrotrophs). Regardless of sub-categories, "parasite" and its long-linked social and religious meanings reveals the back-story of a particular relationship that appears in diverse contexts beyond the bounds of biology.

Equally important but less etymologically telling are the terms used, not for relationships, but for actual creatures: "worms," "*vermes*," or "helminths." The last is the most technical, referring to parasitic worms that inhabit vertebrates, especially in humans. Like "parasite," this ancient word derives from the Greek and was first used in archaic medical writings. Helminths are a key focus in this volume, though in some instances they overlap with appearances of other, less visible parasitic creatures such as bacteria, viruses, even demons. The term "worm" is not narrow in meaning; historically, along with its Latin counterpart, *vermes*, it has indicated bugs, vermin, snakes, and dragons, but most often here, it points to invasive creatures that breach boundaries.

In this volume, worms hold equal importance with parasites because, parasitic or not, they are on the lowest rung of living creatures and perceived as anathema to humanity. Cultural, religious, and even medical meanings often blend their terminology and, by extension, categories, e.g., "parasites" popularly refers to parasitic worms, and not bacteria. This vagueness of classification factors greatly into comparisons because in different contexts and languages, one meaning or signifier may prevail. The inclusion of both terms hopes to cover the phenomenon more comprehensively.

Goals and Scope of this Volume

Persistent and rampant organisms characterized as singularly iniquitous, parasites and worms offer a range of possible research topics: their association with demons, sin, Hell, diets, conceptions of beauty, conceptions of disease, social "parasitism," and social castigation. Such themes appear in many more eras and locales than this one work can address. All the major oral and literary traditions of Egypt, Greece, India, China, Arabia, Persia,

Africa, South America, and North America touch on worms with varying seriousness in regard to religion, medicine, and politics.[4] Our contributors, Gardenour, Weaver and Campbell, Chireau, Baker, More, Vemsani, Crivos, Pitner, Fielding, Thomas, and LeVasseur, proffer a handful of examples, ranging from parasites among the Mbya of Argentina to sinful worms in Hinduism, but the potential for this study supersedes these pages.

The comparative scope of the volume, and the research it wishes to initiate, includes four aspects of human use and experience of worms: physical, psychological, social, and symbolic. The project will investigate their interaction within religious, medical, and cultural traditions, including the scientific tradition. Specifically, this essay discusses biological and biomedical perspectives alongside examples from Chinese and other sources.

The biomedical view provides detailed explications of the physical aspect of worms and parasites, as well as intriguing insights on how parasites can influence human autonomous choice, which lead to the startling possibility that creatures within us may affect cultural and social formation. These scientific angles are especially important because they reveal strong continuities with other cultural vantage points, including that of religion.

Scientific theory may have begun to connect parasites to the social realm, but the former's figurative reach into society has been longstanding. Humans compared to worms or parasites appear cross-culturally in insults, humor, and politics. Generally thought to be undesirable, parasites represent an imbalanced relationship where one entity lives off another; worms are associated with decay, primitivity, and lowliness, antithetical to human values of progress and nobility. The strength of these social metaphors lies in the nature and experience of parasites and worms as "the other," where concrete invasion and corruption of the personal body becomes projected onto the social body.

These metaphors also surface in religion and literature, where parasites and worms are particularly emblematic of death, demons, and destruction. From Nietzsche's "worm of sin" referring to the insidious nature of that concept,[5] to the equation of worms with actual sin in Hinduism, Daoism, and Christianity, et al, the term represents evil and corruption. These symbolic functions deeply inform their social metaphoric use and underline their pervasiveness in cultural domains.

The Power of Disgust

The symbolic, social, and psychological roles of parasites and worms are founded on prevalent responses of disgust and fear. Disgust and impurity have been perennially favored topics among scholars of religion and anthropology. The preeminent example is Mary Douglas, who deduces a common human experience of pollution and dirt.

Her chief argument emerges from a critique of "medical materialism," an approach that explains away traditional cultural and religious practices of purity as proto-scientific forms of hygiene. Douglas counters this claim by excavating a different continuity between modern medical and religious traditions of cleanliness and impurity. In her view, neither religious purity practices nor modern hygiene is built on physical standards of cleanliness. Instead, they are founded on *symbolic* systems intolerant of things that do not fit into their established orders.[6] Consequently, it is the experience of symbolic ambiguity caused by the un-categorizable in both modern hygiene and religious practice, and not questions of mere sanitation, that leads to disgust and its distant cousin, fascination.[7] Moreover, due to symbolic ambiguity's power to threaten the stability of order, it is frequently deployed in myth and poetry, inspiring identification with danger, avoidance, rejection, or even recategorization.

These responses to ambiguity mirror those elicited by parasites and worms as well. Generally shunned, rejected, and perceived as threats, they serve dire purposes in lore and verse literature, and in a striking instance of recategorization detailed in Todd LeVasseur's article, they become beloved, nurtured objects of vermiculture. But while Douglas' theory elucidates parasites' power as ambiguous creatures that threaten human sense of order, it does not go far enough.

Limitations of the symbolic ambiguity theory lie in the false conclusion that symbols are the only link between religious traditions of cleanliness and hygiene. This arises from the assumption that only modern medicine's microscope-empowered vision attributes uncleanness to the presence of organisms like bacteria, viruses, and parasites.[8] In fact, many cultures and religious traditions deem parasites and worms, clearly observable forms of invasion and infection, the very embodiment of violation and contamination and a direct cause of concrete human suffering. This indicates that disgust and un-cleanliness are not just brought about by the disruption of symbolic categories: they derive from the interaction between tangible invasion and pain and cultural and religious values.

Of course, disgust *can* be triggered by socialized symbols, but might it also have ties to "natural symbols," sensory inputs that generate instinctive reactions? In pursuit of these intuitive responses, numerous studies on the potential of disgust's evolutionary functions reveal that people are more repulsed by substances and colors associated with disease and infection. One such study by Val Curtis, et al, surveyed 40,000 people, and the test subjects found a blue viscous substance not nearly as disgusting as a yellow or red one. Furthermore, they judged caterpillars to be much less revolting than *ascaris*, a prevalent parasitic worm. Curtis and company concluded that disgust is an evolutionarily advantageous emotion whose purpose is to help evade infectious disease. This perspective implies that contrary to Douglas' argument, it is not the ambiguity of symbols that causes fear of certain substances, but innate, biological programming.[9]

This sounds akin to the "medical materialism" Douglas critiqued at length, and the evolutionary and instinctive response theory does in fact fall victim to similar imperfections. These studies could easily lead to the riskily reductionist verdict that religious and cultural traditions developed out of this intuitive disgust, even if they lacked the full scope of hygienic knowledge available today. This essay, and the comparative dialogue it hopes to initiate, suggests instead that evolutionary and biological interpretations, though compelling and useful, must not be heedlessly canonized à la the trap of Francis Bacon's Idol of the Cave.[10] Every method is partial. It is possible to employ their varied insights without fully committing to each worldview.

The project thus promotes a plurality of interpretations that bolsters a richer, broader understanding. In the cases of Douglas and the evolutionary school, both proceed from a hermeneutical stance that overrules and dismisses other lines of inquiry. Douglas quells medical materialism just as the medical model topples traditional views of purity. By embracing conversations between culture, symbolic systems, and the interaction of human physiology with the environment, this inclusive, but by no means complete, collection strives for an open discourse and a pooling of knowledge.

Questions of Comparison

In such a wide comparative field, theoretical frameworks play an integral role in gaining entry into a profuse array of particulars. The following is a cross-cultural heuristic outlining some elementary approaches to worms and parasites. Freed from the poles of symbolic and biological systems, this outlook relies on a dialectic between basic perspectives on the nature of the self and other, a foundational relationship epitomized by worms and parasites and the bodies they occupy.

As paradigmatic "others" across cultures and time, worms and parasites solicit two main responses that may be called *oppositional* and *complementary*. The oppositional view presents worms and parasites as radically opposed to the self: the worm is corrupting and evil, an image of chaos and destruction, and something to revile, expel and destroy. In contrast, the latter perspective portrays worms as complementary to humans, and parasites as complementary to their hosts. They are valuable, transformative creatures necessary for survival or the perpetuation of the cycle of life and death. The oppositional and complementary operate dialectically by not essentializing cultures, medicine, or religions. Instead, they represent interpretive angles that, depending on the time and place, move into or out of favor. This non-reductive schema is one way to unearth patterns in human behavior and shifts in cultural systems, allowing for symbolic, medical, and in-between perspectives to interact in a neutral space.

This type of binary relationship analysis matches the theological method in John B. Carman's *Majesty and Meekness*. Carman tackles comparative theology by looking at polarities that frequently populate conceptions of the divine. He invokes the Hindu theology of Ramanuja to discuss how religions attempt to reconcile the contradictory aspects of the divine such as supremacy and accessibility. According to Ramanuja, the polarity of qualities is either understood as complementary or paradoxical/contradictory.[11] Ramanuja argues for the first position, but Carman rightly proposes that his reasoning does not resolve the polarity, and only leads to more conundrums. Carman instead valorizes the paradox created by polar attributes of the divine, like transcendent/imminent or justice/mercy.[12] Such a modus accommodates a lush discussion of the ways different religions approach this contradiction and offers a rubric for comparison.

This method shares similarities with the purposed heuristic. Carman even describes a spectrum of polarities where opposition and complementarity anchor the two extremes.[13] The current heuristic expands on Carman's vision by accentuating the alternation between the two perspectives as more of a dialectic shifting over time. While Carman specifically appreciates the value of polarities in comparative theology, the oppositional and complementary mindsets are viable options for engaging with any type of plurality and field of inquiry, especially those tackling self-other relationships. In the current field, this approach enhances the understanding of human contact with worms and parasites, powerful incarnations of otherness, and how they relate to cultural views on binaries, from divine/human to sickness/body.

Common Helminths in Biomedicine

Helminths, or parasitic worms, are crucial to this study because they embody the relational quality of parasites as well as the physical form of worms. Furthermore, they figure prominently in the intersections between religion and medicine, and in those fields, one can find approaches to healing infectious diseases rooted in both the oppositional and complementary perspectives.

The former sees parasites, including viruses and bacteria, as an invasive enemy force that must be eradicated. The latter subscribes to the ideal of harmonization between worms and humans, where the relationship is properly complementary or mutualistic. Though oppositional and complementary points of view abound in the varieties of medicine and religion, the current biomedical community mainly supports the oppositional perspective. Consequently, creatures commonly identified as helminthes are largely discussed in the context of their propensity to cause disease. Below is a biomedical review of some familiar, observable parasites whose relatives, by virtue of ancient records, are likely to have been spotted in every society, on every continent.

Biomedicine's oppositional opinion of parasites is emphatic in its focus on their material otherness, which is neither spontaneously generated from within the human body nor a form of demonic possession. That is not to say it has no social narrative of its own. In modern medicine, helminths are relegated to "Tropical Medicine," the medicine of "the other," and even labeled as diseases of the "Third World." Actually, they are rife in the West as well, where, amongst other places, they appear in our most beloved "others:" household pets. People annually take their animals for de-worming by the vet; slickly produced television commercials sell treatments to prevent heartworm, a clever animal that can proliferate and prevent the pet's heart from contraction. Yet, it is rare to hear about parasites in "modern" humans. This problem afflicting "others" is in fact not so foreign, and as the data explains below, a closer look at the biomedical angle affirms that parasitic worms are a prosaic part of the human experience, even in "developed" societies, perhaps substantiating the origin of the creatures' common cultural and symbolic functions.

One of the oldest known parasites, confirmed by both archeology and historical records, is the Guinea worm or *dracunculus*. No longer extant outside of East and West Africa, Guinea worms are large and frightening exemplars of what is now synonymous with Tropical Medicine. According to biomedical understanding, humans become infected after ingesting water tainted by microscopic water fleas carrying the worm's larvae. These larvae grow into male and female adults that mate within the host's body. After sexual consummation, the male dies; the female continues to grow and burrows down the host's leg until, after causing a year of acute pain and nausea, it eats through the skin to lay new eggs. The worm encourages its host to spread its seeds by generating an intense burning sensation in the leg, which is somewhat tempered by soaking in water. In this manner, the eggs contaminate a new water source and the cycle begins again.[14] The treatment for this infection has remained unchanged since ancient times. Aside from prevention, the only recourse is to wait in agony for the worm to emerge, and then over weeks or months, extract its entire, unbroken length (which can reach one meter) by slowly wrapping it around a stick.

Biomedicine affords a detailed account of the groups, physical forms, and symptoms of various creatures that invade the body. *Dracunculus* is but one type of nematode (roundworm), a large category of parasites much more widespread than realized. The diseases they cause affect not only poverty-stricken individuals in the developing or remote areas of the world, but also people elsewhere, including the United States.

Ascaris, or roundworm of the intestines, is found throughout the world in both temperate and tropical regions where sanitation and hygiene are poor.

Every person in those areas may be harboring the *ascaris,* which afflicts an estimated 1.4 billion people. The illness is most common in children between the ages of 3 and 8, and the World Health Organization has recorded approximately 60,000 annual deaths worldwide.[15]

A less devastating but even more commonplace roundworm is the pinworm, *enterobius*. In the US alone, the Center for Disease Control finds that it affects up to 50% of people under 18 and those who care for the infected. This particular worm lives in the anus and ventures out at night to lay eggs in the hopes of reproduction. Infection does not cause significant symptoms other than itching, though this problem can lead to insomnia.[16]

Like the WHO and CDC, the Food and Drug Administration deals with parasites as well. Another roundworm, the Seal worm *anasakis*, is transmitted via un-cooked, under-cooked, or cold-smoked salmon. The FDA considers all wild Pacific salmon hosts to this parasite, and has discovered especially high levels in wild Atlantic salmon. This infection is rarely diagnosed in the U.S., but considering the popularity of sushi, one might suppose many mild cases go untreated. *Anasakis* infection is easily misidentified as acute appendicitis, Crohn's disease, gastric ulcer, or gastrointestinal cancer, and often resists detection till the patient coughs up the offending worm.[17]

The tapeworm is the largest and perhaps the most terrifying of all. Thriving in intestines, it can grow to many meters in length and cause great debility. The two most troubling types are the beef tapeworm *Taenia saginata* (5-17 meters) and the pork tapeworm *Taenia solium* (3-8 meters), contracted from eating tainted, raw, or insufficiently cooked meats. Unlike most parasites, which evolved to live in other animals and leap to humans by accident, these worms have always been at home in people. Partially due to this, they can live up to 30 years, longer than most pets. With an estimated 50 million cases a year each, both parasites are widely distributed.[18]

Taeniasis, an infection of the intestines by adult tapeworms, can cause abdominal pain, weight loss, digestive disturbances, and possible intestinal obstruction. The more serious tapeworm disease is cysticercosis, where cysts grow on various organs. An individual with cysticercocis involving the central nervous system (neurocysticercosis) may suffer from psychiatric disorders or epileptic seizures. In such instances, death is common.[19]

The number of helminths far exceeds these few examples, and includes assorted flukes and flatworms. These select cases offer a glimpse into the biomedical knowledge of parasitic worms, attest to the pervasiveness of these creatures, and highlight their visibly invasive presence even prior to the invention of the microscope.

Complementary Biomedicine?

In an historic shift away from the once dominant oppositional view, biomedicine has begun to embrace a complementary approach, Unwittingly illustrating the periodic tendency for heuristic alternation, this change first began with the small parasites called bacteria, once considered dangerous without exception. Now, even mass-market advertisements tout the "good" bacteria in yogurt that aid digestion. The move toward a more complementary vision signals a veering away from unchecked, antibiotic carpet-bombing of all "invaders." It seems that many of the "invaders" are actually natives. This thawing of medical dogma may have also granted helminths clemency.

A new field of research presumes that human bodies have evolved in concert with manifold parasites, and that in ultra-sanitized conditions, people no longer house the proper array of their little companions. This theory asserts that the human immune system expects to find parasites, but when its expectations are unmet, confusion ensues. The researchers postulate that this is the true instigator of immune-mediated afflictions like inflammatory bowel disease, asthma, multiple sclerosis, and autoimmune diabetes.

Investigators like David Elliot and Joel Weinstock explore this premise through experimental therapies that repair the imbalance by reintroducing worms to calm the immune system,[20] which in turn alleviates symptoms of these autoimmune diseases.[21] In light of the age-old, negative reception of worms, selling such a therapy to the public would likely encounter resistance. However, it may be that as with bacteria there are good and bad kinds of worms, and this negative impulse, as a defensive instincts designed to avoid ingesting the bad kind, can be overcome with the ascendency of a complementary viewpoint.

Other Medical Perspectives on Worms

Biomedicine provides one medical perspective on worms and disease, but many others exist. They often occupy a reality where the medical, religious, and symbolic are not airtight categories. In these traditions we find both the oppositional and complementary perspectives with demons and worms as infectious disease, or parasitic worms as a result of internal imbalance.

Chinese medicine, a medical tradition pigeonholed as holistic and the counter to biomedicine's dominant oppositional approach of isolation and destruction, has witnessed a long history of parasitic treatment. The following is an account of the nine worms associated with both physical forms of disease and demonic forces. The great physician Sun Simiao (580-681?) explains:

> The hidden worm is the master of the group. The "round" worm pierces people's hearts and kills them. The white worm reproduces itself, and each generation increases in number. The mother increases in size, sometimes reaching four or five *cun* and can be lethal.[22] The flesh worm makes people vexed and arrogant. The lungworm causes people to cough while the stomach worm causes people to vomit and belch. The weakening worm, also called the diaphragm worm, makes people spit. The red worm makes people's intestines rumble. The "pin" worm lives between the outer body and the intestines. If they are numerous, they cause piles. If they are fierce, they cause skin problems (leprosy). Depending on people's wounds, they create various sores, shallow and deep, favus, psoriasis, scabies, and tooth decay.[23]

Detailing the creatures' ill effects and visible attributes, this record could be interpreted as proto-scientific, but its sense of worms, with their variety, specific sizes, and symptoms exists within a separate reality. These worms were seen as infectious creatures, as in biomedicine, but they were also viewed as manifestations of bodily imbalance or semi-demonic harbingers of death. These two different valuations, one oppositional and the other complementary, showcase radically different approaches to illness.

The writings of Wang Kentang (1549-1613) exhibit the first approach. Allying worms with the external cycles of nature, he believes they originate in moisture and heat, and blames their acquisition on a diet of "warming" and "damp" food like excess meat, alcohol, fish, and soft shell turtle. Though biomedicine could claim most of these are common parasitic vectors, the worms' association with warmth and dampness integrates them into a system where these causes are also symptoms. When the body becomes too hot and damp it manifests worms that cause symptoms grouped with those qualities of imbalance. In this case, heat and damp specifically affect the spleen, understood as the ruler of digestion, mucous, flesh, and vitality. These functions of the spleen correlate to the symptoms of worm infection, thumping heart and abdominal pain, vomiting foamy saliva, sallow gaunt face, dark circles under the eyes, little appetite and inability to gain weight, silent, desiring sleep, and easily hot or cold.[24] These parasites operate in a worldview not ruled by opposed entities, but by the balance of complementary forces of hot and cold, damp and dry, and an interconnected system of organs.

This naturalistic and complementary system of parasitic medicine was not the only Chinese cultural response to the nine worms. The *Yutang Dafa* (*The Great Rites of the Jade Hall*), a 12th century Daoist text by Lu Shizhong, presents an intricate vision of parasitic disease as demonic invasion.[25] This disease does not result from natural imbalance but from worms as supernatural contagion. Lu describes this illness in detail, including six distinct stages of "worm" infection. For a cure, instead of relying on herbs (associated with natural patterns like hot and cold), he advises the use of incantations and talismans to command gods, otherworldly authorities, to expel the demonic worms.[26]

This form of medicine takes the oppositional view of parasites as unwanted interlopers, a stand surprisingly close to that of biomedicine, which sees disease as intrusion and corruption that must be driven out. Where the Daoist position diverges is in the blurring of Western boundaries of the natural and supernatural. If the disease is more serious, the demonic forces become more concrete and one sees worms during purgation; in minor cases, the confirmation of successful exorcism/healing is simply a dream of a person's departure. This overlap of physical and spiritual also relates to the numerous traditions that associate demonic worms/parasites with mental or moral afflictions, including Sun Simiao and his vexation and anger causing flesh worm.

Whether affecting mind or body, religious and medical traditions do not always distain worms as evil disease bringers, for there also exist instances where they are portrayed as complementary and necessary destructive forces. Examples include the work of celebrated Islamic philosopher and physician Ibn Sina or Avicenna (10th-11th c.) who thought digestion required the activity of worms.[27] As this bodily function requires breaking things down, he logically assigns this role to Nature's specialist.

Chinese Buddhists (circa 5th-7th c.) proponed that there are eighty thousand worms essential to life in the body.[28] These worms are complements, or more accurately, components that reinforce the Buddhist view of the body as a rotting bag of flesh. One of the writings using this image, the *Fayuan zhulin,* presents this wormy vision in a meditation practice intended to achieve the ideal of No-Self or *annatā*. Once one realizes the body is rife with defiled creatures, it and the self can be seen as just a composite of ephemeral, smaller beings. This may seem like a condemnation of worms, but they in fact are symbols of Truth that point to the reality of impermanence.

Moreover, worms have been used in a healing capacity, though generally as a homeopathic, magical, or "poison"-type treatment. One Chinese *Materia Medica* entry prescribes for an inflamed eyelid the use of a roundworm vomited by a child after it has been dried, powdered, and made into a paste. In Europe, parasites were also used as medicine, either dried or burned, as anthelmintics.[29] In this latter case, parasites are still agents of disease, but they can cure, too—corruption can be used to corrupt the corruptors.

These positive uses concur with the work of Elliot and Weinstock, who believe that human flourishing is not defined by total purity, but by integration into a world-ecosystem. Macrocosmically within this system and microcosmically within the human body, there is a relationship between all parts, large and small. They need to be properly balanced, e.g., a healthy immune system needs a certain amount of interaction with helminths. However, there is a distinction between these "good" parasites, and a preponderance of their kind that is malignant and should be eliminated.

The overlap and interplay of the two needed views of oppositional and complementary approaches to medicine have existed in their various forms throughout history, including examples like the Chinese or biomedical ones. Their dynamic shifting will continue in the endless movements of medicine's search for total health.

Worms, mind, and morality

Regardless of one's approach to parasites and worms, their infection is often thought to alter the mind as well as the body. Variant perspectives about parasites and their effects on the mind expose a tangle of questions about individual identity and free will. Are parasites complements or do they overpower and manipulate? On this issue, scientific and other cultural or religious views tend toward an oppositional attitude, perhaps due to the tendency to links parasites and worms to mental conditions like misbehavior, immorality, or possession by evil.

One of the most studied parasites in biomedicine that modifies psychology is *Taxoplasma gondii or* T. gondii. Evolutionarily, it is designed to encourage transmission to its native host, cats. When rats are infected, it alters their brain chemistry so that they become attracted to the smell of cat urine instead of being fearful of it. Such a change raises the likelihood of predation and consumption by a cat, which completes T. gondii's journey back to its preferred species. Research concludes that this parasite evolved to control animals in a particular way, and behavioral changes are not just side effects of infection. This raises the possibility that human volition might also be radically altered and ultimately controlled by parasites.[30]

Other studies of T. gondii, examine its specific effects on human behavior. Most conclusively, research correlates it with schizophrenia: higher levels of distinct antibodies in pregnant women correspond to higher rates of schizophrenia in their children.[31] That T. gondii infects people via cat excrement also provides a pop-psych explanation for the proverbial "crazy cat lady" who despite others' objections and escalating odors remains blissfully unperturbed by overflowing litter boxes.

Though this one parasite is a particularly salient manipulator of the human mind, innumerable others are likely to exist, having yet to be understood. To take it a step further, if these critters can partially drive human behavior, could there be a legal, substantiatable claim for parasite-induced insanity in the future? What would this mean to established beliefs of human agency, and by extension, standards of morality? In fact, religious thinkers have long regarded parasites and worms as negative moral influences.

Prominent among them is Ge Hong (283-343), a Chinese alchemist. His theory of the Three Corpse Worms highlights the immoral and demonic aspects of parasites.[32] Strictly oppositional in his views, Ge indicts corpse worms as the depraved agents driving people to indulge base desires: greed, gluttony, and lust. These worms also play the role of cosmic moral punisher as they provide all the fruits of misbehavior in the forms of suffering, pain, illness, and death.[33] Thus, for Ge Hong and alchemists seeking immortality, purging these entities is a prerequisite to surmounting death and its associated sins.[34]

Ge's identification of parasites as psychological manipulators and instigators of immoral behavior has strong parallels to the biomedical belief that these organisms affect the human mind in addition to the body. Though these perspectives arise from divergent contexts, they raise the same profound questions about human identity and the origins of evil. The confluent views of science and religion both claim that parasites, demons, or a mixture of both, can possess and control humans. But interestingly, while in some religions these invaders are understood as just physical worms, the worms of science could be interpreted as demonic, for they too possess and manipulate. What, then, is the boundary between physical worms influencing the mind and demonic possession? Such a distinction remains blurred in many cultural arenas.

Self, mind, and culture

Established correlations between parasite infection and human behavior suggest that parasites could also influence the collective behaviors that comprise society and culture.

The foremost authority on this subject is Kevin Lafferty of the U.S. Geological Survey who normally investigates parasites as measures of ecosystem health. This background inspired him to explore the dynamic between environment, parasite behavior modification of humans, and cultural patterns. First, he notes high infection rates often follow humans' transformation of their own habitat, activity that might result from parasites influencing behavior for their own benefit. Second, since all cultures developed in environments harboring parasites, Lafferty proposes that one could use high instances of specific parasite infestations to quantitatively analyze aggregate personalities of groups in the same way sociologists employ categories like sex, class, or individualism to examine patterns of behavior.[35]

Over all, he considers parasites' behavioral influences a positive, a complementary force, because they engender greater cultural diversity, in contrast to pundits like Eric Jones, discussed below, who singles out parasites for impeding human progress. In his preliminary look at manipulation, Lafferty turns to the aforementioned T. gondii, because its effects on humans are thoroughly researched and infects nearly half the world's population. He looks for correspondence between high rates of the infection with particular group behaviors. This is possible due to T. gondii's culpability in both radical conditions like schizophrenia and milder habit modifications. Its influences vary by sex. Afflicted women tend toward an increase in intelligence and the strength of the super-ego, causing greater dutifulness, rule-consciousness, and attentiveness to others. Infected men experience the opposite transformation: a decrease in intelligence and super-ego, leading to greater frugality, rigidity, and loyalty. Both groups become prone to guilt and neuroticism.[36] The changes in behavior result mostly from heightened dopamine levels, which encourage risk-taking and underlie the daring of rats discussed earlier.[37]

Accordingly, Lafferty inquires whether a greater degree of sex role differentiation, rampant neuroticism, or desire for control and structure exist in locations with higher rates of infection. Unfortunately, the data does not definitively prove this hypothesis. While the connection is confirmed in Western countries, it is much less evident in Asia or Africa.[38] This finding hints that T. gondii is just one of many elements that shape culture, an expected result for students of history.

Such a preliminary study does not collapse traditional notions of cultural creation, but it does open up an intriguing line of inquiry. It has long been accepted that the environment partly shapes cultural variation, but perhaps parasites are an internalized aspect of its influence, less easily noticed than external factors like weather and terrain. Now, scholars can broach a whole new set of previously unconsidered questions.

In the future, parasitologists and humanities scholars will hopefully produce fruitful collaborations in a potentially new field. Using a parasitological approach to cultural analysis reintroduces a serious issue that has troubled humanity since the beginning of the Modern Age: is there a dividing line between humanity and nature, and if so, where does it lie?

Much of anthropology has been devoted to exploring how culture differs from nature, but if one considers that parasites and their workings cross this chasm, this division begins to fracture. If parasites significantly influence human behavior, then can we say civilization is solely a human creation? What if it is more the result of collaboration where parasites are possibly a driving, creative force, especially behind the alteration of environments underlying the foundation of society? This may inaugurate a more integrated understanding of nature and culture, as well as a more complementary comprehension of parasites. On the other hand, these organisms might have been a miniscule influence on human history, and are just along for the ride. Either way, this topic deserves attention in clarifying the relationship with our intimate "other."

Worms, Parasites, and the Social Other

Lafferty's theory posits that hidden biological entities can partly shape cultural divisions. This approach, while valuable, does not explain the full role of parasites and worms in society. A more comprehensive vision would include the social metaphor of humans as parasitic or worm-like, as well as controversial claims that often apply such language to people "corrupted" by actual parasites.

Parasites and worms are perfect symbols for the social "other." Their identification with vilified people or groups appears throughout disparate cultural contexts. The instinctively negative response elicited by these creatures is used to demonize, dehumanize, debase, or more subtly offer social critique and visions of imbalanced relationships. Frequently and passionately, people label others as "worms" or "parasites" to condemn them as the destructive and worthless parts of society.

In *The European Miracle*, Eric L. Jones combines this symbolic equation of worms to the socially inferior "other" with actual parasitic infection. The incendiary Jones strongly believes that infections in other societies inform why Europe became the dominant world power. He cites mind-boggling statistics: the total weight of liver parasites in China circa 1987 was equal to that of two million people; in 1960, 90% of Chinese had tapeworms, and 30% died from helminth related issues. Such data from East Asia, he argues, partially explain European supremacy because without the disruptive cycles of parasitic debility, European civilization was comparatively more stable and productive.[39] Maintaining such a biased thesis requires selective readings of history: Jones conveniently forgets the incredible creative and organizational power demonstrated by the Chinese civilization, including the building of massive public works uninhibited by the apparent epidemic.

Aside from pointing out Jones' willful blindness, there are other ways to critique his assumptions that place the weight of dominant-submissive dynamics of entire populations on a small class of organisms. First, they call to mind the colonial rhetoric of legitimation that promotes conquest as a duty to purify and sanitize the backward peoples of the world. Second, Jones' theory implies a pseudo-evolutionary outlook. The farther the distance (the less one is contaminated) from the un-evolved worm, the more advanced a civilization must be. More specifically, this plays into a narrative dating to, at least, Hegel's *Philosophy of History*, an influential work propounding that the teleological path of humanity moves from East to West. Hegel identifies China as the least advanced society in the Far East and Germany as the pinnacle of human achievement in the West.[40] Jones fuses this Orientalist outlook with the symbolism of lowly parasites in his description of the extreme other as the worm-ridden East.

The relationships of parasites to group variation can be a legitimate field of study without arguing positions where "those people" are infected and so inferior; in this regard, Lafferty's scholarship is refreshingly progressive. Still, the danger remains in conflating worms and parasites as the physical other with their symbolic representation in the social other.

Revisiting China, one finds lucid ancient examples of the pure social metaphor. Shang Yang (390-338 BCE) and Hanfeizi (280-233 BCE) in their political writings use worms or bugs to represent people who drain communal or political strength. Shang Yang describes the six lice among officials who waste resources and undermine the government: those who help the elderly or the unemployed, and those who seek beauty, love, success, or virtuous conduct for themselves.[41] Hanfeizi echoes this view with his discussion of the five gnawing worms: Confucians who praise the sage kings, sophists, soldiers of fortune, rich merchants, and selfish officials.[42] Both of these social critiques clearly communicate the authors' values and castigating views.

Other deployments of this social metaphor abound, even in the etymology of the word "parasite" itself. Charlotte Baker's article touches on this topic, but first consider a few cases. During war or genocide, comparisons to parasites and worms are often used to dehumanize the enemy. World War II saw such tactics on both sides of the conflict. As early as *Mein Kampf,* Hitler cursed the Jews as an infection, a cancer, but perhaps most potently as a parasite draining the rest of Germany.[43] This is the ultimate metaphor for all manner of unwanted outsiders, framing them as deadly bloodsuckers on the social body, filthy creatures of the dirt. These associations coincide with the Jews' identification as *Untermensch* (sub-humans), the opposite of Hitler's ideal *Ubermensch* (superhuman).

In Nazi Germany, following the brutal logic of the metaphor, as with parasites of the body, parasites of society like the Jews and other outsiders could only be dealt with through purgation and total removal.[44] Though painful to recall, this case demonstrates the power of the metaphor, arguably enhanced by the unconscious dread of actual parasites. Hopefully, future studies of politicized rhetoric can note the visceral origins of symbols like parasites.

Hitler's trenchant and destructive imagery was not unique during the war. The Allies also used worms to dehumanize, though not quite as gruesomely. One British war bond poster depicts Hitler's head on an earthworm being torn from the ground by a Union Jack-sporting bird. Though the symbolic connotations of the earthworm are not quite as vile as those evoked by helminths, the creature's weakness, lowliness, dirtiness, and inhumanity are unmistakable.

British media again employed this propagandist technique on 20 February 2003. The cover of a Paris edition of *The Sun*, a tabloid from the United Kingdom, planted Jacques Chirac's head on the body of a worm eating through France. This image was in response to Chirac's unwillingness to support his allies, America and Britain, in invading Iraq. The accompanying article described him as "spineless," a term closely related to the negative characterization of worms in English.

Concurrent with their social metaphoric usage, parasites and worms inspire hate and fear. Yet, this is a fear that we are not so different from the others: vile, lowly, filthy, weak, selfish, and mindless in our consumption of resources that may not even belong to us. As J. Z. Smith has indicated, the other is actually a reflection of the self. The social use of parasites to impute supremacy over an infected, debilitated, and parasitic other works by redirecting self-reflection. Demonization of others in general and specifically of the parasitic and wormy kind is a powerful way to remove guilt over violence or failure. Perhaps the social application of parasitic symbolism and language is a form of psychological displacement to cope with people's own worm-worthy behavior. After all, parasites are an apt analogy for the invasive and exploitative colonialists who generated Jones' "European Miracle."

Whatever else one can say about colonialism, it was a profitable venture that garnered much envy and respect, especially from other aspiring powers. A famous early American parasitologist, Maurice C. Hall, noted the irony in society's veneration of certain kinds of people who live off the work of others. With their positions advantageously founded on an imbalanced economic exchange, the already rich and powerful acquire a disproportional concentration of wealth and influence, the most favored markers of social worth, until they become the most successful of parasites, who are not only experts of exploitation, but are considered all the more inspirational for it.[45]

The interplay between the simultaneous idolization and rejection of social parasitism appears in two popular American political narratives. One condemns the wealthy, who purportedly profit off the efforts of others, building capital by scheming money away from hard working people. The other narrative indicts the poor who live off welfare and the "Nanny State."[46] In this latter vision, the government is actually the primary parasite, taxing away resources to support its bloated, useless body. By siding with those who occupy higher social and economic echelons, this second narrative implicitly differentiates between deserving and undeserving parasites.

In all cases, emotional rhetoric based on the metaphor motivates people to political action, another illustration of the potency behind the symbols' close association with disgust. To whom or what people ascribe these terms reveals the values of their society and demonstrates the ways that responses of disgust can be formulated and manipulated within disparate cultural contexts: at times, the same image can be similarly employed to further opposite political agendas.

The Symbolic Power of Worms and Parasites

Much social imagery of parasites and worms correlates to those found in religious and cultural symbolic traditions. Such symbolism tends to be oppositional and not complementary, since these entities recur as harbingers or embodiments of death, corruption, and destruction. As mentioned before, visceral instincts may partly explain the force of these connections, but cultural symbols and uses also likely derive from observations of dead trees, animals, and human corpses being consumed by creepy crawly things. This kind of experience semiotically binds these creatures and their consumptive activities to mortality and decay. Some cultures take this vision to an extreme: in the traditional Czech view, the body itself was thought to be composed entirely of worms, which at death finally returns to their true state.[47] More casually, slang terms in English like "worm condo" refer to cadavers.

As worms coincide with corruption and rot, they are frequently grouped with Hell, demons, filth, evil, and ruin. On the human level, worms represent immorality, sin, desire, base animal passions (the need to consume and possess), and transgressions of proper social boundaries. They are insidious and inescapable, an inner influence beyond conscious self-control. Cosmically, worms and parasites tend to stand for the hidden, the undifferentiated, the interior corruption of all things, and the decay of chaos. As Sartre said, "Nothingness lies coiled in the Heart of Being—like a worm."[48] In this simile, Nothingness is worm-like in that it penetrates and undermines Being by defiling its seeming perfection and wholeness from within. As masters of the cycle of life and death, worms can symbolize the primordial, that to which all things revert. But as living representations of change and impermanence, they are the opposite of the divine, the eternal, the immortal, and the pure.

The preceding discussion generalizes the symbolic significance of worms and parasites and such archetypal language threatens a foundational concept of semiotics. Much ink has been spilt in demonstrating the arbitrary nature of symbols and signs, and undoubtedly an ever-present arbitrary component exists in all human cultural creations. The connection of worms to death *is* arbitrary in the sense that it is an arbitration, or choice, but it is also based on a common or natural sense of meaning and experience. Worms, even on a symbolic level, stir up innate, existential fear, which suggests that cultural symbols are at least partially emotional/instinctual categories, and not purely intellectual ones.[49] The questions then become how fully is the sign motivated and where is the boundary between the symbol's natural meaning and the arbitrary meaning ascribed by cultural context and intellectual creation? The articles in this volume afford a glimmer of the answers, which are a moving target, but a worthwhile one.

Those who doubt this or the evolutionarily minded study mentioned previously would be advised to undertake a simple experiment. Search YouTube for "bot flies," "stomach worms," or any type of parasite already mentioned. While watching the video clips, carefully observe any mental, emotional, or physical responses to the images. Quite likely, viewing these creatures will set off a negative visceral reaction akin to fear and disgust. Arguments like that of Douglas, which pinpoint the origin of disgust in conceptual ambiguity, do not explain such effects. These are sense-mediated reactions not based on abstract ideas or culturally particular rules of cleanliness. Yet, while these feelings are shared across diverse human circumstances, this instinct can be augmented or diminished by other cultural analytic factors. For this reason, claims of symbolic universality for something like parasites remain untenable. Instead, consider that certain signs tightly bound to instinctual responses have special power with a greater portion of "meaning." This may be why symbols like parasites can move entire nations to violent action, or prevent people from even opening this book.

Complementary and Oppositional

While oppositional views inspired by disgust and fear dominate in cultural arenas, the symbolic framework of worms and parasites lends the influence of instinctive responses to complementary perspectives as well. For example, worms connection to change need not solely signify destruction; from a complementary angle, it might connote necessary, even positive, transformation. This type of dual meaning is thoughtfully addressed in Allison More's essay.

Such binary understanding of worms raises subtle questions about the nature of good and evil. One can argue that cross-culturally, conceptions historically shift between two views: destructive forces are part of the cycle of life (complementary); they are an enemy to guard against and annihilate (oppositional). Though the aforementioned polarity between the human and the divine solicit the same theoretical approaches, the reactions to parasites or worms match their lines of reasoning precisely. In fact, one might even ask parallel theological questions: Does God flow within humans (complementary)? Are parasites complements to humans created by God? Is God hopelessly removed from this corrupted world (oppositional)? Are parasites creatures of the Devil eternally opposed to humans? The repercussions in choosing either pair of options are manifold, affecting one's fundamental view of the cosmos. It changes approaches to healing; it alters conceptions of evil; it transforms humanity's role in the world.

An illustration of this theological conundrum centers on an ectoparasite, lice, which was previously thought to be inseparable from humans. Lacking lice in Medieval Europe was tantamount to an absence of humanity; lepers, shunned for their moribund condition, never contracted lice. In this context, theologians debated whether Adam and Eve had lice before the Fall, and if lice and other parasites traveled on Noah's Ark.[50] These questions may sound strange to modern ears, but they deal with the nature of evil and suffering. If creatures causing such distress predate sin, then suffering is a natural state of creation (complementary); if not, then such suffering and even the existence of parasites result from sin (oppositional). Either view affects the perception of the human-divine bond, but neither emerges as an indisputable conclusion.

Complementary and oppositional views must co-exist because they each represent an essential aspect of the self-other relationship. To use a theological example, consider the outcome of an extreme oppositional view: if God were completely transcendent, such a being would have no meaning to humanity. There must be some affinity or there is no relationship at all. Similarly, extreme complementarity results in a lack of distinctiveness between self and other, and collapses the relationship as well. In the case of parasites, they must have the potential for positive symbolic meaning in addition to negative connotations in order to function as an effective "other" that is neither too dissociated nor too similar to humanity.

In summation, these two views work dialectically, both in terms of how they interact and how each engages experience; they seem to alternate in and out of favor throughout history in a range of systems. This heuristic does not attempt to explain away religious or scientific methodologies, but instead suggest another approach that emphasizes the powerful, continual interplay of complementary and oppositional engagements with the world.

Conclusion

This introductory essay initiates a discussion, continued in the articles that follow, and hopefully carried on beyond the confines of this book. In short, parasites and worms, lowly as they may be, deserve more attention. They hold a notable place in our development as a species, and the study of these unassuming creatures will not only help to demystify a common phenomenon, but also further understanding of myriad civilizations.

In this chapter, parasites have been used to broach questions about the boundary between culture and nature, the derivative issue of "natural" symbols, the inflammatory employment of visceral symbols in political rhetoric, the suspect divisions between religion and medicine, and elementary ways to contextualize the self, the other, and the world. In addition to laying out these themes and defining and furnishing principal examples of worms and parasites, this article has also offered a simple analytical tool. The heuristic of oppositional and complementary perspectives is one among many, but it affords an encompassing approach to religion, medicine, and politics.

The following essays, with their eclectic range of topics, comprise a preliminary attempt to understand parasites in culture and history. Within them, one discovers fine analyses of a shared human experience and the rich array of creative responses it has engendered across continents and centuries.

Acknowledgements

Many contributed either directly or indirectly to the fruition of this project. I must thank Shigehisa Kuriyama for first encouraging me to pursue this odd line of research, my wonderful co-editor Brenda Gardenour for her feedback and especially for inspiring the conference for which I first composed this paper, Linda Barnes for her invaluable help in reframing the early drafts, my Lulu whose constant support and nuanced sense of language were invaluable, and lastly, my Mother who inspired me to see meaning in all things great and small.

Notes

1. For a fine treatment of this topic, see J. Z. Smith's irreverent use of parasites as the perfect other to disparage a pretentious discussion of "the other." He both employs parasites because of their lowly status for his critique, while also quite seriously demonstrating their usefulness as a metaphor of the other. Jonathan Z. Smith, "What a Difference a Difference Makes," in *Relating Religion* (Chicago: U Chicago Press, 2004): 145-159.
2. Alfred C. Reed, "Animal Parasites of Man and Their Control," *Science* 72, no. 1877 (Dec. 17, 1930): 611.
3. For a overview of the categories of symbiosis, their historical development, and relationship to the social sense of the terms see, Jan Sapp, "On the Origin of Symbiosis" in J. *Symbiosis and Stress,* ed. Joseph Seckbach and Martin Grube (New York: Springer, 2010), 5-18.
4. For the best overview of this raw material to date, though presented via an old history of science methodology, see R. Hoeppli, *Parasites and Parasitic Infections in Early Medicine and Science* (Singapore: University of Malaya Press, 1959).
5. Friedrich Nietzsche, *The Anti-Christ,* trans. H. L. Mencken (Tuscon, AZ: Sharp Press, 1999), 90.
6. Mary Douglas, *Purity and Danger: An Analysis of the Concept of Pollution and Taboo* (London: Routledge and Kegan Paul, 1966), 35.
7. *Ibid.*, 37.
8. *Ibid.*, 35.
9. Val Curtis, Robert Aunger and Tamer Rabie, "Evidence that Disgust Evolved to Protect from Risk of Disease," *Proceedings of the Royal Society B* 271 (2004): S131-S133.
10. One of Bacon's Four Idols, the idols of the Cave refers to humanity's tendency to take one sort of knowledge and haphazardly apply it to everything. The term for this intellectual fallacy references Plato, but also implies the trap of thinking that every "cave" resembles one's own. See Francis Bacon, *Novum Organum*, trans. Peter Urbach and John Gibson (Peru, Il: Open Court Publishing, 1994).
11. John B. Carman, *Majesty and Meekness: A Comparative Study of the Contrast and Harmony in the Concept of God* (Grand Rapids, MI: Wm. B. Eerdmans Publishing, 1995), 12.
12. *Ibid.*, 13.
13. *Ibid.*, 15.
14. On a more religious note, the "fiery serpent" referenced in the Hebrew Bible is sometimes understood as the Guinea worm. Num. 21:4-9.
15. "Ascariasis," accessed July 1, 2011, http://www.who.int/water_sanitation_health/diseases/ascariasis/en/.
16. "Enterobiasis," accessed July 1, 2011, http://www.cdc.gov/parasites/pinworm/
17. "Anisakis simplex," accessed July 1, 2011, http://www.fda.gov/Food/FoodSafety/FoodborneIllness/FoodborneIllnessFoodbornePathogensNaturalToxins/BadBugBook/ucm070768.htm.

18. "Parasites–Taeniasis," accessed July 1, 2011: http://www.cdc.gov/parasites/taeniasis/disease.html.
19. Christopher M. DeGiorgio, M.D et al, "Neurocysticercosis," *Epilepsy Currents* 4, no. 3 (May 2004): 107–111.
20. This represents a more thoughtful use of helminthic therapy, in contrast to the modern tapeworm diet. This diet, like its 19th century counterpart, involves ingesting tapeworm eggs to help one lose weight. After attaining one's desired weight, one takes an anthelmintic to purge the parasites. This type of treatment therapy raises philosophical questions about human desire and issues of consumption's relationship to control. This phenomenon permits human gluttony, a parasitic quality, by employing worms to consume one's excess consumption. The result is an outer appearance concordant with genuine self-control, thereby maintaining the currently accepted physical markers of self-discipline. See http://www.tapewormdiet.net/.
21. D.E. Elliot and J.V. Weinstock, "Helminthic Therapy: Using Worms to Treat Immune-mediated Disease," *Advances in Experimental Medicine and Biology* 666 (2009): 157-66.
22. *Cun* is a Chinese measurement equivalent to the length of the middle bone of the index finger (used in acupuncture) or about 1.31 inches.
23. *Gujin Tushu Jicheng*, vol. 450 (Shanghai: Zhong hua shu ju, 1934), 22.
24. *Ibid.*
25. The meaning of the text has been the subject of debate, especially whether it is an empirical record of the Tuberculosis bacteria. See Liu Ts'un-yan, "Taoists' Knowledge of Tuberculosis in the Twelfth Century," *T'oung Pao* 57, no. 5 (1971): 285-301; Michel Strickman, *Chinese Magical Medicine* (Stanford: Stanford University Press, 2002), 34-38.
26. *Daozang* (Taibei: Yiwen yinshuguan, 1977), fasc. 220.
27. *Ibid.*, 164.
28. *Foshuo ershi toutuojing* 佛說十二頭陀經, T17.721a19; *Fayuan zhulin* 法苑珠林, T53.902a2.
29. Hoeppli, 168.
30. M. Berdoy, et al, "Fatal Attraction in Rats with *Toxoplasma gondii*," *Proceedings: Biological Sciences* 267, no. 1452 (August 7, 2000): 1591-1594.
31. Preben Bo Mortensen, et al, "Early Infections of *Toxoplasma gondii* and the Later Development of Schizophrenia," *Schizophrenia Bulletin* 33, no. 3 (May 2007): 741-744.
32. For more on this theory see, Livia Kohn, "Koshin: A Taoist Cult in Japan Part II: Historical Development," *Japanese Religions* 20, no. 1 (Jan 1995): 123-142.
33. This connection of behavior, punishment, and worms also speaks to the effect of behavior on worm infestation. Though not always conceived in a moral framework, actions that cause one to be "impure" often lead to parasite infection and illness. The strongest example of this phenomenon is *dracunculus*, a parasite that may be the first human disease eradicated solely through behavioral modification. This hypothesis is also explored in the article by Weaver and Campbell.
34. *Ibid.*, 77.

35. Kevin D. Lafferty, "Can the Common Brain Parasite, *Toxoplasma gondii,* Influence Human Culture?," *Proceedings of the Royal Society B* 273 n. 1602 (Nov. 7, 2006): 2749.
36. Lafferty, 2749.
37. *Ibid.*, 2750.
38. *Ibid.*, 2754.
39. Eric L. Jones, *The European Miracle* (Cambridge: Cambridge University Press, 1987), 6.
40. Georg Wilhelm Friedri Hegel, *Philosophy of History,* trans. J. Sibree (New York: Prometheus Books, 1990).
41. Shang, Yang, *The Book of Lord Shang*, trans. J. J. L. Duyvendak (Clark, NJ: The Lawbook Exchange, Ltd., 2003), 256.
42. Han Fei, *Han Fei Tzu: Basic Writings,* trans. Burton Watson (New York: Columbia University Press, 1964), 96.
43. Adolf Hitler, *Mein Kampf,* (London: Hurst and Blackett, 1939), 261-262.
44. Also see, G. G. Otto, *Der Jude als Weltparasit* [The Jew as World Parasite] (Munich: Eher Verlag, 1943).
45. Maurice C. Hall, "The Glorification of Parasitism," *The Scientific Monthly* 33, no. 1 (July 1931): 42-52.
46. A related view appears in Ayn Rand, *The Fountainhead* (New York: Signet, 1996), 665.
47. Josef Cizmar, *Lidove Ledarsivi v. Ceskoslovensku.* 2 vols. (Brno: Melantrich, A.S. 1947), 261, quoted in Frances Cattermole-Tally, "The Intrusion of Animals into the Human Body: Fantasy and Reality," *Folklore* 106 (1995): 90.
48. Jean-Paul Sartre, *Being and Nothingness,* Hazel E. Barnes trans. (New York: Citadel Press, 2001), 21.
49. Worms in particular suit a discussion on the interplay of instinct and intellect, feeling and thought; they are the opposite of the clean, orderly, divided, intellectual, and technological. Through their extreme polarity, these associations elucidate the common use of parasites in science fiction. In that context, the contrast is greatest when inarticulate worms, the lowliest and least evolved of all creatures, control advanced humans of the future. As one will see in Julian Fielding's article, *Star Trek, Alien,* amongst others, abound with variations on this theme.
50. R. Hoeppli, *Parasites and Parasitic Infections in Early Medicine and Science* (Singapore: University of Malaya Press, 1959), 401.

Serpents and Sanitation: A Biocultural Survey of Snake Worship, Cultural Adaptation, and Parasitic Disease in Ancient and Modern India

Lesley Jo Weaver, Amber R Campbell Hibbs

The idea that humans adapt both biologically and behaviorally to their environments has a long history in anthropology and other schools of thought. The concept of *cultural adaptation* suggests that humans respond to challenges posed by their environments in such forms as disease, food shortage, and natural disaster by adapting not only biologically, but also culturally. Cultural practices are a useful way of "speeding up" the process of adaptation, since group behaviors can be changed much more quickly than genetic makeup, which requires generations of natural selection to change. By allowing people to respond rapidly to environmental challenges, cultural adaptation improves group survivorship. Since its inception, cultural adaptation has undergone multiple revisions and contestations from both anthropological and biological theorists, and yet it remains a useful way of thinking about humans' recursive relationships with their environments.[1]

A classic, and controversial, example of cultural adaptation is the widespread taboo on pork consumption, which is observed with striking frequency in Jewish and Muslim communities around the world. Some scholars have speculated that pork taboos may be so widespread because avoidance of pork reduces incidence of trichinosis, a highly damaging and occasionally deadly parasite that is spread exclusively through consumption of undercooked pork.[2] It is easy to comprehend how any practice that reduced pork consumption would likewise reduce the chances of contracting trichinosis and, over time, result in a healthier overall population with better group-level survivorship.

Naga (serpent-deity) worship in North India may be an instance of cultural adaptation to environmental stress induced by high incidence of parasitic infection. Naga worship is an important religious practice in contemporary India that dates back thousands of years. It is particularly common in the northern mountainous regions of the country, where nagas are often understood to be the patron deities of villages.[3] Because it is believed that nagas reside in water sources like pools, ponds, and rivers, one important expression of naga worship is respect of water sources. This behavior may result in particularly careful avoidance of defecation near water that is used for drinking and cooking. By discouraging defecation near water sources, belief in nagas could therefore reduce the prevalence of parasitic infections spread through the fecal-oral route. Naga worship may be "adaptive" not in the sense that people would knowingly use it to control parasitic disease, but in the sense that it may have positive secondary consequences that result from modified defecation behavior.

Cultural Adaptation, Taboos, and Human Health

Scholars have long debated whether or not some underlying logic motivates cultural taboos. A prototypic example of such controversy was instigated in the Middle Ages by Jewish philosopher and doctor Maimonides, who was the first to justify the rabbinic code of *kashrut* (kosher) against pork consumption on secular, rather than religious or moral, grounds. Because of the pig's unclean habits, pork was believed to be difficult to digest and generally bad for health. Maimonides therefore suggested that the pork taboo had become so widespread because Jewish religious officials had historically used it to coerce the faithful into avoiding its harmful effects on health.[4] When American anatomist Joseph Leidy discovered in the mid-1840s that the parasitic infection trichinosis was contracted through consumption of undercooked pork, some scholars heralded this discovery as corroboration of Maimonides' theory, and as affirmation of the "logical" reasoning that they presumed was the underpinning of this and other religious codes. Such foresight on the part of rabbinic officials in a time when trichinosis—let alone germ theory—had yet to be discovered is unlikely, however, and to interpret pork sanctions as a purposive public health measure may be erroneous.

Pig taboos resurface with striking regularity across cultural contexts and religious orientations, and so this debate has continued.[5] Recognizing the improbability of there being a conscious, premeditated rationale for pig taboos, contemporary anthropologists Mary Douglas and Marvin Harris proposed alternative explanations. Douglas suggested that the taboo was a response to the taxonomic quandary posed by the pig, the only cloven-hoofed *non*-cud-chewing animal which would have been known to ancient rabbinic and biblical scholars. As it does not fit neatly into the *kashrut* classification of animals that are permissible to eat (that is, cud-chewing animals with cloven hooves), Douglas hypothesized that the pig represented "matter-out-of-place," and she suggested that such un-classifiability would have been deeply offensive to the human psyche, particularly in the biblical societies, which relied heavily upon hierarchy and classification to maintain order and make sense of the world.[6] Thus the pig taboo was a response to this matter-out-of-place, a way to maintain the important cosmological order of *kashrut* by completely excluding that which did not conform. Harris, on the other hand, suggested that the prohibition was related to the more pragmatic, materialist consideration of the fact that raising pigs in the climates of ancient Middle East would require extraordinary effort, environmental stress, and personal resource drain (and in short, would not make economic sense). Others suggest that taboos—especially those surrounding meat—may indeed serve a health-protecting purpose, since meat is the most likely of all food types to carry pathogens.[7] On the other hand, plenty of studies of food taboos have concluded that they are *harmful*, at least at the individual level, because they restrict usable food sources in environments where food may already be limited.[8]

The key question in the study of this and other taboos is whether or not certain avoidances are a good adaptive strategy in certain cultural contexts, and whether or not they may have real health benefits for those who observe them. This discussion is grounded in notions of cultural adaptation, a wing of scholarship that was incorporated on a large scale into the social sciences in the late 1950s, when Frank Livingstone suggested that sickle cell anemia in West Africa was an adaptive response to endemic malaria there (1958).[9] Following this seminal paper, a group of studies investigated the ways in which human genetics and behavior patterns fluctuate in response to the challenges posed by environmental (usually disease- or poverty-related) stressors.

Anthropologist Alexander Alland played an important role in the popularization of adaptation theory in cultural studies (1969, 1975). In evolutionary biology, the relative adaptiveness of a trait has often been assessed by its effects on an organism's fertility, fecundity, morbidity, and/or mortality; but Alland suggested that the adaptability of cultural traits be assessed by their ability to endure over time. This selective retention by cultural transmission between generations was therefore the mechanism originally proposed for the spread of certain cultural traits, in a process analogous to but not synonymous with a central tenet of natural selection; namely, that organisms with a novel advantageous genetic trait have increased likelihood of passing it on to their offspring as a result of enhanced reproductive fitness.[10] Assuming that, like genetic mutations, cultural traits arise by chance, those traits that have been preserved from one generation to the next might serve some valuable emotional or utilitarian function, and for this reason would be transmitted to offspring.[11] What remains to be determined are the specific mechanisms by which an individual chooses to retain or discard a given cultural practice.[12]

As with all theoretical orientations, there are some caveats that go along with a perspective based on cultural adaptation. An obvious tautology arises if we assume that all extant cultural traits must be adaptive simply by virtue of their existence in the present. Alland addressed this problem as follows:

> To say that adaptive traits are those which are present in systems, or that those traits which are present in systems are adaptive, adds nothing to our understanding of process. To be meaningful, adaptation as a temporal process of transgenerational change must have some kind of independent measure and/or be charted according to a consistent theory (1975).

Likewise, Brown cautions:

> The concept of adaptation does not imply that the resulting biological or cultural traits are the only solution to a specific environmental problem, nor does it imply that they are the optimal solutions. Rather, adaptation means that the...traits are tolerable or have minimally sufficient positive consequences to improve an organism's chances for survival in a particular environment (1981).

Thus, a cultural adaptationist perspective asserts that "adaptive" traits arise not as purposive responses to an individual's or a group's particular needs, but rather as random occurrences whose preservation has associated trade-offs *and* positive secondary consequences, be they in terms of increasing group size, energy availability, inclusive fitness, group health, or a host of other possibilities. Herein lies an important difference between evolutionary adaptation and cultural adaptation: evolutionary adaptation occurs at the genetic level, whereas cultural adaptation exerts its influence not at the individual or even the kin group level, but upon larger groups in a long-term process of "cultural group selection."[13]

In his work on anthropology and epidemiology in Singapore, Fred Dunn developed a conceptual model to clarify this important distinction between health-related behaviors that are deliberate as opposed to those that are not.[14] Dunn, who advocated strongly for the inclusion of behavioral factors in epidemiologic studies, lays out a schema by which health-enhancing and health-harming behaviors are classified into subcategories of deliberate non-deliberate actions. He classifies health-related behaviors that are not explicitly cultivated for health enhancement but which nevertheless have salutary health effects as "non-deliberate health enhancing."[15] People who participate in a certain health-enhancing practice may have better survivorship than their neighbors who do not adhere to this practice, and thus may live longer and in larger numbers and have more opportunity to pass this practice along to future generations. It is not necessary that people be aware of the benefit of a practice, so long as its benefit is potent enough to affect their actual survivorship. This distinction responds to the assumption that cultural adaptations are the result of "conscious decision-making processes, or even subjective awareness of the 'problems' that diseases pose for those who suffer from them."[16] Although it has been the target of significant criticism and revision, when applied non-reductionistically, the concept of adaptation has proven again and again to be a useful heuristic tool for thinking about the relationships between humans, culture, and their environments.[17]

If we apply this logic to our pork taboo example, it becomes apparent that the persistence of the pork prohibition over time is not sufficient grounds to conclude that those who instated it knew it would reduce health problems associated with pork consumption, as Maimonides originally suggested. That the pork taboo has survived and spread when many other religious practices have not may instead indicate that this prohibition has some utility to the groups who practice it, including the potential health benefit of avoiding trichinosis, a highly pathogenic parasite that is spread exclusively through the consumption of undercooked pork. This unforeseen (and, likely, unconscious) utility could have given individuals who adhere to the pork taboo better health, and therefore longer lives and more opportunity to teach the behavior to subsequent generations, than their pork-eating counterparts. However, the benefit may also have been unrelated directly to health; for instance, it could have been an ecological (Harris) or taxonomic (Douglas) benefit. Ultimately, we can only conclude that the widespread and longstanding persistence of the pork taboo may suggest some underlying logic for its preservation, even though this utility may not be directly understood by those who practice it.

In recent decades, an approach based on cultural adaptation has been used to examine the associations between disease-induced stress on populations and the adaptiveness of culture in response to such stressors. This wing of scholarship is replete with studies on parasitic diseases, such as filariasis in Malayan aborigines, as studied by Dunn in 1972, malaria in Sardinia, as studied by Brown in 1981, and *Ascaris lumbricoides* in Ethiopia as studied by Vecchiato in 1997.[18] Other studies examine nutrition, poverty, and disease in Peru[19], infant feeding practices and survival, and blood pressure and cultural consonance.[20] All of these studies, and many more, have explored how culture shapes people's exposures and responses to disease stress, often in unexpected or unintended ways.

Naga Worship in India

We suggest that naga worship may have important health-protective effects which have contributed to the longevity of the practice. Nagas are serpentine deities found primarily in the Dharmic traditions, usually depicted as a physical and psychic blend of human and snake-like characteristics.[21] They are understood to be the controllers of terrestrial and celestial waters, possessors and conduits of vital liquids such as semen, sap, and poison, protectors of the earth's material and magical riches, and guardians of specific territories.[22] They are holy, powerful, and wrathful beings who are believed to bestow enhanced fertility and earthly riches upon those who please and propitiate them, and curse those who offend them with infertility, skin boils, poor harvests, and poverty. The shadowy realm in which they reside, called *nagaloka*, is normally invisible to humans but is usually understood to be in or near a water source.[23] As a folk belief system, naga worship falls outside the formal boundaries of codified religion, but aspects of it have been incorporated into modern Hinduism and Buddhism.[24] Because they are nature-bound deities, nagas are believed to interact closely and regularly with humans, and as such are more accessible for propitiation than some of the other divine beings in the Hindu and Buddhist pantheons.[25]

Naga worship can be traced textually back to the Vedic period (1500 to 400 B.C.E.) when Buddhism and Hinduism had not yet been differentiated from one another. The late Vedic *Satapatha Brahmana* states: "'To those snakes who are in the shining sky, or those who are in the rays of the sun, or those whose home is in the water. Honor to those snakes!' Wherever they are, this honors them. One does this by saying, '*nama, nama*' for honor is sacrifice."[26] Although they share a common origin in the Vedas, nagas are depicted somewhat differently in later Hindu and Buddhist texts. In earlier Hindu texts, they are usually portrayed as half-human, half-snake creatures with fiery tempers.[27] In the later Buddhist texts, however, nagas tend to be purely serpentine, less anthropomorphic beings. Some authors speculate that this shift was a conscious effort on the part of early Buddhist redactors to un-deify nagas, thereby preventing them from competing with the Buddha for devotees' spiritual attentions.[28] Despite efforts to quell naga devotion, some degree of belief in nagas is has survived in people who self-identify as Buddhist and Hindu today.[29]

In both traditions, the nagaloka underworld maintains an association with water sources, and as such, pools, wells, and rivers often serve as foci for *naga pooja* (worship). Out of respect for nagas' holiness and fear of their volatility, actions that are considered impure are strictly prohibited in naga realms, and one of the most polluting actions in the Vedic purity-pollution system is defecation.[30] Therefore, observance of the naga cult would theoretically limit defecation around water sources, and indeed beliefs about the polluting nature of bodily substances *do* mitigate behaviors (*Ibid.).* To borrow from Dunn's conceptual framework, naga worship thus construed would fall into the realm of "non-deliberate health enhancing" behaviors: naga worship would be adaptive because it has positive health consequences, causing people to sequester defecation away from water sources and thus reducing fecal contamination of water sources.[31]

Passages that explicitly associate nagas with water sources also abound in ancient Sanskrit and Pali literature, which represent some of the oldest religious texts worldwide that are still in use. Naga worship is not mentioned in the *Rig Veda*, the earliest of the four Vedas, but it is fully developed by the time of the Sanskrit epic *Mahabharata*, composed and expanded between the 5th century B.C.E. and 4th century C.E.[32] According to the *Adi Parvan*, the introductory section of the *Mahabharata*, the entire world is held up by the naga Sesha, the "world-serpent," who carries the earth on his head and cradles it in his endless coils, and is also often depicted in later sources with the Hindu god Vishnu.[33] The *Adi Parvan* contains other vivid descriptions that highlight nagas' power, ubiquity, and association with water, such as the following: "The snakes whose king is Airavata shine in battle, and are like rain clouds streaked with lightning and driven by the wind. Beautiful and ugly with speckled coils, the ones who arose from Airavata shine like the sun in the vault of the sky. The Nagas have many paths on the north bank of the Ganges."[34] Nagas' associations with specific geographic features (in this case, the Ganges river) become increasingly important as we examine later sources, including contemporary naga worship.

Cities that are situated on the bank of the Ganges, such as Varanasi, have historically been important sites of naga worship. In Varanasi, there are special sites dedicated for the purpose, including two ornately-carved *kuan* (wells) in which nagas are believed to reside.[35] Polluting actions of any sort, including not only defecation, but spitting, urinating, or entering the water without bathing one's feet first, are strictly prohibited (Cozad 2004, personal observation). Varanasi also serves as the setting for many of the early Buddhist Jataka Tales, fable-like Sanskrit and Pali poems dated somewhere between 300 B.C.E. and 400 C.E. that describe the Buddha's journey through previous lifetimes.[36] In these tales, nagas make multiple and often colorful appearances. The *Ghatasana Jataka*, for example, contains an especially vivid warning of the consequences that may ensue when nagas' waters are polluted:

> Once upon a time a flock of birds had taken up their abode in a large tree which stretched its leafy branches over the waters of a lake. All these birds, roosting in the boughs, let their droppings fall into the waters below. Now that lake was the abode of Chanda, the Naga king, who, being wroth at this fouling of the water, resolved to take vengeance on the birds and burn them out. So one night when they were all roosting along the branches, he set to work to drive them away: first he made the waters of the lake to boil, then he caused smoke to arise, and thirdly he made flames dart up as high as a palm-tree.[37]

Another Jataka, the *Campeyya Jataka*, explicitly links the physical wellbeing of its naga protagonist to the state of his home waters. Upon leaving for a journey, the Naga king Campeyya tells his queen Sumana, "If anyone should strike me, good woman, the water in this pond will become dirty. If the *suparnas* [garuda birds] should seize me, the water will go away. And if a snake harmer should take hold of me, the water will appear blood-red".[38] Here, the naga is so intimately tied with his watery home that it reflects the state of his physical being, and if he is harmed, the water too becomes sullied. Thus, dirty waters may not only incite nagas' wrath, but they may also indicate that the resident naga is in physical peril.

The early Hindu *Skanda Purana* (probably composed shortly after the close of the Vedic period, 4th–5th century C.E.), emphasizes the importance of maintaining cleanliness around water much more directly than do the Jataka tales; it instructs that one's feet must be washed before entering any water source for bathing, and enjoins readers to abstain from spitting, throwing dirt, urinating, and defecating in water.[39] Taken together, the Jataka tales and the peri-Vedic injunctions provide the foundations for the forbearance that naga believers observe around water sources today.

Indeed, naga worship is an important religious practice across contemporary India, and the locality of the folk system remains strong.[40] In the northern mountainous regions, nagas are often nominally and historically linked with specific villages near which they are believed to reside[41]. Legends of familial, usually sibling, relations between patron nagas of proximate villages may shape inter-village relations and marriage patterns.[42] Additionally, the festival Naga Panchami, which falls on the fifth day of the moonlit-fortnight in the lunar month Shravan (July/August), is widely celebrated in the northern states of Uttar Pradesh and Bihar, West Bengal, and Maharashtra, and also across South India. The festival is first mentioned in the *Skanda Purana,* where, as paraphrased by Kumar (1983), it describes how "the great *nagas*…built nine *kundas* [pools] which became famous *tirthas* [crossings] bestowing all virtues. This place became famous as Svetadvipa, where Lord Hari [Krishna] is reclining on Sesanaga [Sesha, the world-serpent]. Here Nagas should be worshipped on *pancami* of *Sravana.*"[43] During this festival, people propitiate naga statues and live snakes with milk, ghee, water, and rice. They also ritually bathe at specific pools, wells, ponds, and/or sections of river that are associated with nagas and nagaloka. People abstain from plowing fields in order to avoid inadvertently disturbing or harming snakes, which, even during non-festival times, are often treated as embodiments of naga spirits.[44] Naga worship has even found its way into recent popular media, such as Harper's magazine.[45]

Naga Worship: A Culturally Adaptive Practice?

Naga worship influences people's relations with water sources in significant and potentially beneficial ways: It may mitigate waterborne disease prevalence by discouraging defecation, spitting, urinating, and the discarding dirt into water that is used for drinking. Particularly because of nagas' associations with local hydrologic features, one might expect waterborne parasite prevalences at the local level to be negatively associated with prevalence of naga worship. Unfortunately, a satisfactory regional comparison of defecation behaviors (particularly as they relate to religious observance) and prevalence of parasites has yet to be conducted in the Indian Himalaya, but the literature affords tentative support for the hypothesis that naga worship may be linked to decreased prevalence of parasitic infection.

Broadly speaking, there is strong evidence that parasitic infections are widely prevalent in India, and that behavior modifications may impact upon this prevalence. Helminth and protozoal infections are well-documented and significant health problems in both urban and rural India, including high-altitude and relatively remote mountainous regions.[46] Furthermore, several studies have successfully identified risk factors for parasitic disease associated with defecation practices. "Promiscuous defecation," defined as defecating anywhere within village premises as opposed to defecating in the woods outside of a village, is one such risk factor, as is lack of facilities for feces disposal.[47] Other studies have shown that population parasite load is associated with behavioral patterns around water, including two studies that examine schistosomiasis transmission in Nigeria and Kenya in connection with water usage and frequency of contact with water sources.[48] Increased contact with contaminated water sources, as expected, results in higher levels of infection. Parasite infection is a multi-causal phenomenon, but poor personal hygiene, sanitation practices, and local water quality are well-established risk factors.

Additionally, there is tentative evidence that areas where naga worship occurs may have lower prevalence of certain parasites, although no causal association has been made. Two limited studies in the Himachal Pradesh and Kalimpong areas of north India report low levels of parasitic—particularly protozoal—infection.[49] Observes one author:

> It is rather surprising that the Himalayan hill people in spite of very unsatisfactory sanitary arrangements show such low incidence of protozoal infections, not much

> higher than Sweden, England, Northwest Germany and college freshmen in the United States. ...How far the ritual ablutions, demanded especially by the Hindu religion, are strictly observed I have not been able to ascertain, but a scrupulous cleaning of the hand that is used for eating would certainly diminish the risk of acquiring infections.[50]

There may be a connection between ritual observance and reduced incidence of parasitic infection, although it should be noted that it is quite possible that this reduction is due to the high altitude of the regions rather than to behavioral proscriptions around defecation. Other scholars working in the Kalimpong and broader Himachal Pradesh areas, however, have noted that naga worship is widespread there, to the extent that special temples are often constructed to honor them.[51] Most of the villages dispersed throughout this sparsely populated, rugged landscape are close to a water source, usually a glacial lake or spring, and naga worship often takes place on or near the water's edge. (Based on personal observation. See Figure 1)

Figure 1: A small mountaintop temple in the Parvati Valley, Himachal Pradesh, India. Note the serpentine figure running across the top of the roof. (Source: Authors, 2005)

These sources indicate simultaneous low prevalence of protozoal infection in areas of active naga worship. Although it would be overly presumptuous to make conclusions about the potential health utility of naga belief with such limited data, these preliminary studies suggest intriguing future directions for an investigation of the associations between religious practice and parasitic infection in mountainous areas of India.

Conclusions

The remarkable persistence of naga worship from Vedic times, through the historic split of Hinduism and Buddhism, and into the present may be due, at least in part, to its limiting of defecation near water sources, thereby reducing fecal contamination and parasitic infections, and increasing overall group health. This position, borrowed from cultural adaptation theory, suggests that naga-worshiping groups may have a health-related "edge" over groups who do not worship nagas because naga worship discourages defecation near water sources. Any practice that reduces fecal contamination of water is likely to have some benefit for human health. We suggest that if this benefit was strong enough, it could have historically resulted in higher survivorship in the naga-worshiping groups, thus giving them more opportunity to pass along the practice of naga worship to future generations. The salutary effects of cleaner water would, ultimately, contribute to the conservation of naga worship through generations and into the present.

There are undoubtedly many reasons for the continuation of naga worship from ancient times into the present aside from its potential health benefits. The cultural value placed on traditional beliefs and customs is, in and of itself, a significant motivator for the inter-generational transmission of practices like naga worship. Our rather functionalist explanation for the persistence of naga worship does not deny or preclude the important role of tradition in promoting these practices over time. In fact, tradition is a crucial part of adaptation theory: Without tradition to motivate people's transmission of beliefs from one generation to the next, cultural adaptation would not be possible.

The vast majority of recent studies that have examined the adaptiveness of cultural traits to disease stress (such as Brown 1981, Dunn 1972, and Vecchiato 1997) have had access to reliable epidemiological data in the areas where they have worked, which provides evidence that certain practices actually *do* change the incidence or prevalence of a particular condition.

There is, of course, no way to compare parasite prevalence between naga-worshiping and non-naga-worshiping groups without direct observation and measurement. Rigorous fieldwork comparing the regional prevalence of intestinal parasitism in areas where nagas are worshipped and in similar areas where they are not could begin to address this question. At present, the potential long-term benefits to human health remain one potential, and compelling, explanation for the persistence of naga worship in India.

Acknowledgement

"This material is based upon work supported under a National Science Foundation Graduate Research Fellowship. Any opinions, findings, conclusions or recommendations expressed in this publication are those of the author(s) and do not necessarily reflect the views of the National Science Foundation."

Notes

1. See A. Alland, "Ecology and adaptation to parasitic diseases," in *Environment and Cultural Behavior: ecological Studies in Cultural Anthropology*, ed. AP Vayda (New York: Natural History Press, 1969); A. Alland, "Adaptation." *Annual Review of Anthropology* 4 (1975):59–73; FL Dunn and CR Janes, "Introduction: Medical anthropology and epidemiology," in *Anthropology and Epidemiology*, eds. CR Janes, R Stall, SM Gifford (Boston: D. Reidel Publishing Company, 1986); SJ Gould and RC Lewontin, "The spandrels of San Marco and the Panglossian Paradigm: A critique of the adaptationist programme." *Proceedings of the Royal Society of London B* 205.1161(1979):581–598; M. Singer, "Farewell to Adaptationism: Unnatural selection and the politics of biology," *Medical Anthropology Quarterly* 10.4 (1996):496–515; AH Goodman and TL Leatherman, "Traversing the chasm between biology and culture: An introduction," in *Building a new biocultural synthesis: political-economic perspectives on human biology* (Ann Arbor: University of Michigan Press, 1998)
2. *The guide of the perplexed by Moses Maimonides*, trans. S Pines (Chicago: University of Chicago Press, 1963); M. Douglas, *Purity and Danger* (London: Routledge, 2000); M. Harris, *Good to Eat: Riddles of Food and Culture.* (Long Grove, IL: Waveland Press, 1998); OC Handa, *Naga Cults and Traditions in the Western Himalay.* (New Delhi: Indus Publishing Company, 2004)
3. Handa *Naga Cults*, 2004; L. Cozad, *Sacred Snakes: Orthodox Images of Indian Snake Worship* (Aurora, CO: Davies Group, 2004); JPH Vogel, *Indian Serpent-lore* (Hertford: Stephen Austin & Sons, Ltd, 1926)

4. Maimonides, *Guide to the Perplexed*
5. N. Henrich, J, Henrich, "The evolution of cultural adaptations and how it created pregnancy and lactation food taboos that protect against marine toxins"(Unpublished)
6. Douglas, *Purity and Danger*, 2000
7. Henrich and Henrich unpublished; DMT Fessler and CD Navarette, "Meat is good to taboo: Dietary proscriptions as a product of the interaction of psychological mechanisms and social processes," *Journal of Cognition and Culture* 3.1(2003):1–40; Fessler, DMT. "Reproductive immunosuppression and diet." *Current Anthropology* 43.1(2002): 19–61
8. J. Katona-Apte, "The socio-cultural aspects of food avoidance in a low-income population in Tamilnad, South India." *Journal of Tropical Pediatrics and Environmental Child Health* 23.2(1977):83–90; O. Obgeide, "Nutritional hazards of food taboos and preferences in Mid-West Nigeria." *American Journal of Clinical Nutrition* 27(1974): 213–216; R. Aunger, "Are food avoidances maladaptive in the Ituri forest of Zaire?" *Journal of Anthropological Research* 50.3(1994):277–310; MI Santos-Torres and Vásquez-Garibay, E. "Food taboos among nursing mothers of Mexico." *Journal of Health, Population, and Nutrition* 21.2(2003):142–149
9. FB Livingstone, "Anthropological implications of sickle cell gene distribution in West Africa." *American Anthropologist* 60.3(1958), 533–562
10. WH Durham, "The Adaptive Significance of Cultural Behavior." *Human Ecology* 4.2(1976):89–121
11. Alland, *Adaptation*, 1975
12. Durham, "The Adaptive Significance," 1976
13. J. Stoltis, R. Boyd, PJ Richerson, "Can group-functional behaviors evolve by cultural group selection? An empirical test." *Current Anthropology* 36.3(1995):473–494
14. Dunn, "Introduction," 1986
15. *Ibid.*
16. P Brown, "Cultural adaptations to endemic malaria: problems of comparison." *Human Ecology* 14.3(1986):311–332
17. Goodman and Leatherman, *Traversing the Chasm*, 1998; , Singer, "Farewell," 1996; E. Mayr, "How to carry out the adaptationist program?" *The American Naturalist* 121.3(1983):324–334
18. FL Dunn, "Intestinal parasitism in Malayan aborigines." *Bulletin of the World Health Organization* 46(1972), 99–113; P Brown, "Cultural adaptations to endemic malaria in Sardinia." *Medical Anthropology* 5.3(1981):313–339; N Vecchiato, "'Digestive worms': ethnomedical approaches to intestinal parasitism in Southern Ethiopia." In: *The anthropology of infectious disease: International health perspectives.* C Inhorn and PJ Brown, eds. (Australia: Gordon and Breach, 1997)
19. T. Leatherman, "A biocultural perspective on health and household economy in southern Peru." *Medical Anthropology Quarterly* 10.4(1996):476–495; TW McDade and CM Worthman, "The weanling's dilemma reconsidered: A biocultural analysis of breastfeeding ecology." *Developmental and Behavioral Pediatrics* 19.4(1998):287–299.

20. W Dressler and JR Bindon, "The health consequences of cultural consonance: cultural dimensions of lifestyle, social support, and arterial blood pressure in an African American community." *American Anthropologist* 102.2(2000):244–260
21. Vogel, *Indian Serpent-lore*, 1926
22. LW Bloss, *Ancient Indian Folk Religion as Seen through the Symbolism of the Naga* [dissertation].(Chicago: University of Chicago, 1971)
23. Cozad, *Sacred Snakes,* 2004; Vogel, *Indian Serpent-lore*, 1926; S Kumar, *The Pauranic lore of holy water-places.* (New Delhi: Munshiram Manoharlal Publishers Pvt. Ltd, 1983)
24. Bloss, *Ancient Indian Folk Religion*, 1971
25. *Ibid.*; Cozad 2004; Handa 2004
26. *Satapatha Brahmana*, 7.4.1.28–30, cited in Cozad, 2004, 28–29
27. It is believed that the Taoic fire-breathing dragon is a derivation of the *naga*, transmitted back to China by the Buddhist monk Hiuen Tsiang during his visit to India sometime between 629–45 C.E. (Vogel 1926).
28. Cozad 2004; Handa 2004
29. Vogel 1926, Handa 2004
30. L Dumont, *Homo hierarchicus: The caste system and its implications* (Chicago: University of Chicago Press, 1974)
31. Dunn, 1986
32. Vogel, 1926
33. Vogel, 1926
34. *Adi Parvan* 1.3.139–144, cited in Cozad 2004, 53.
35. Vogel, 1926; Cozad, 2004
36. The Jataka tales also contain some apocryphal additions as recent as the 19th Century: Ashliman, DL. *The Jataka Tales.* 2002. Accessed 8 March 2008. http://www.pitt.edu/~dash/jataka.html
37. *Ghatasana Jataka*, cited in Vogel 1926, 136
38. *Campeyya Jataka*, 10, 17–22, cited in Cozad 2004, 97
39. *Skanda Purana*, I–2.64.20–24, cited in Kumar 1983, 13
40. Bloss, 1971
41. Handa 2004; Cozad 2004; Vogel, 1926
42. Vogel, 1926; personal observation, 2005
43. *Skanda Purana* 5–1.65.23, cited in Kumar 1983, 128
44. Cozad, 2004
45. M Berlinski, "Woman marries snake: a peculiar Indian love story." *Harper's Magazine* 315.1890(2007), 41–52.
46. P Singh, ML Gupta, TS Thakur, NK Vaidya, "Intestinal parasitism in Himachal Pradesh." *Indian Journal of Medical Sciences* 45.8(1991):201–204; KP Jithendran, "Protozoan diseases of livestock, wild animals, and man in Himachal Pradesh: An overview." *ENVIS Bulletin: Himalayan Ecology and Development* 10.2(2002):4–10; RJ Traub, "The role of dogs in transmission of gastrointestinal parasites in a remote tea-

growing community in northeastern India." *American Journal of Tropical Medicine and Hygiene* 67.5(2002), 539–545.

47. Nishiura et. al., 2002; Kar, 2000; Rao et. al., 2001; Cifuentes et. al., 2000
48. Akogun and Akogun 1996, Kloos et al 1997
49. Svensson 1956, Singh et. al., 1991
50. Svensson 1956, 166
51. Handa 2004, 125–30

WORMS AND THE CORPORAL BODY IN INDIA: AN EXAMINATION OF HINDU LITERARY TRADITIONS

Levanya Vemsani

A Pox on Thee!

All Indian traditions, Hinduism, Jainism and Buddhism[1] view worms[2] as the basest of living creatures. This extreme status of worms provides a fascinating angle to investigate the interrelation of humans and other creatures in both Hinduism and traditional Indian medicine, the *Āyurveda*. This paper undertakes a historical analysis of the relationship of worms, body, and soul to better understand the socio-administrative ritual place of these lowly worms in the worldly environment, as well as the cosmos.

Worms do not just hold a place as the basest creatures, but also relate to past sins, as their defiling influence degrades and encumbers a soul's true divine self; one can either be reborn as a worm due to one's past sins or can be infested with them. This paper will examine in detail the opposition of worms and humans in religion, ritual and administration. I, furthermore, intend to explore the nature and function of Hindu narratives associated with worms and what they reveal about the opposition of worms and humans. I will also examine the existence of worms in other worldly contexts, such as heavens, hells or underworld in relation to their existence on earth.

Although worms and the human body represent completely opposite ends of the spectrum of living creatures, there are certain circumstances when they might cohabit, giving rise to the "wormy corpus," an altogether different entity. When and how does the purest and highest creature in the world, the human body, become vulnerable and possessed by the lowest of the life forms on earth, the worms? What are the philosophical and theological reasons and implications for such invasion? In order to examine these questions, I will begin my survey of the worms and the body with an analysis of the myths from Hindu texts regarding the origin of worms and the body, in the second main section below, which will also shed light on the socio-religious views on worms and the body.

What happens when a wormy corpus exists in this world? What is its status? What rules govern its social, religious and political roles? In order to find answers for these questions, in the third main section, I will study the official and medical documents (medical and administrative texts as well as official donative inscriptions from temples) for their value in understanding the socio-religious practices with regards to the wormy corpus.

This then leads to the next question: How does one understand the existence of the body, the worms, as well as the wormy corpus in not only this world, but also the other worlds of Hindu cosmology? I will address this issue, in the fourth main section of this paper, and discuss the connection between worms, the body and never ending cycle of life between these different realities. I conclude the results of my investigation of the worms, and body in the final section.

These five sections will draw on an array of texts from the Hindu literary tradition.[3] I begin with references in the Vedic tradition and examine the epics (the *Mahābhārata* (M.bh.) and the *Rāmāyaṇa*), and also the *purāṇas*. I will furthermore utilize the *āyurvedic* medical texts of India, the *Charakasaṃhita* and *Suśrutasaṃhita,* to understand and compare medical views on worms to literary and religious ones. As all these texts date between 1500 B.C.E–1000 C.E., I will also examine the contemporary inscriptions and judicial and administrative texts including *Manudharmaśāstra* (Manu) and *Arthaśāstra* (Artha), covering a range of time periods in order to understand the variety of views on the worms and the body. *Manudharmaśāstra*, dated between 100–300 C.E., includes numerous prescriptions that address the ritualized actions prompted by disease or interaction with worms. These include religious as well as social practices of first millennium India (100–1000 C.E).

My examination of worms and the body in religious, inscriptional, and medical texts aims to elucidate the nature and function of worms as understood in Indian culture, Indian images of the body, as well as Indian notions of sin and punishment in the administrative, judicial, and medicinal contexts.

Worms, Disease and the Body in Hinduism: Thought

Worms in Hinduism represent more than a biological entity, they are imbued with cultural, social and religious meanings. These cultural understandings depict the human body and worms as opposite entities. The body is depicted as superior and pure, while the worms are depicted as lowly and impure. Here, worms occupy the bottom level of the circle of life (*saṃsāra*).[4] Because of these notions, the relationship of worms to humans reflects on topics of purity, ritual, moral and ethical wellbeing, and social stratification. Furthermore, the conception of the diseased body relates to the conceptions of worms and body. These two issues combine in the following summary of myths that discuss origins of worms, the body, and wormy corpus, the relationship of disease causing organisms (worms) to sin, and the cosmological role of the worms, all of which provide a foundation for further discussion.

Creation of Worms and the Body: *Manudharmaśāstra* I. 12–40

Everything in this world is created by Brahmā, the creator, including the worms. I include below a brief summary of the creation myth:

> The divine one resided in that egg during the whole year, then he himself by his thought (alone) divided it into two halves. And out of those two halves he formed heaven and earth, between them the middle sphere, the eight points of the horizon, and the eternal abode of the waters. From himself, he also created the mind, the five objects of sensation, and joining the minute particles of those six, he created all beings. Because those six minute particles enter the creatures, the frame is called *śarīra* (body). Moreover, in order to distinguish actions, he separated merit from demerit, and he caused the creatures to be affected by the pairs (of opposites) such as pain and pleasure. Small and large worms and beetles, moths, lice, flies, bugs, all stinging and biting insects and several kinds of immovable things are thus created.

This myth has Brahmā creating the body (*śarīra*) by bringing together the mind and the five objects of sensation (speech, sight, touch, smell and hearing) that originated from himself. As the body (*śarīra*) thus created is subject to actions due to the possession of five objects of sensation, he distinguished good or bad actions with the mechanism of their resulting pain or pleasure. Because moral choices are clarified by affects on the body, this myth presents morality and the body as intrinsically connected. Furthermore, the body in this narrative represents good, the lesser things such as the worms, moths can be interpreted to represent the bad side of actions and its associated pain, something highlighted by the emphasis on their stinging and biting nature. This myth reveals a foundational connection of the sinful actions of the body with the negative effects, the pain and suffering, caused by worms. Therefore, in this mythology the body and worms are eternally connected, as the body is always subject to action and thus to the pain caused by worms.

This relationship is further clarified in another Vedic myth that explains the creation of human body as a pure weapon (corporal body), which only later became infused with sin to distinguish it from the divine, and it was this detail that made humans vulnerable to the punishment of worms as disease.

> The gods created a thunderbolt to attack demons, and that was man (*purusha*). They hurled it at the demons, and it scattered the demons, but then it turned back to the gods. They were afraid of it and took it and divided into three pieces. They put sin[5] in him: sleep, laziness, anger, hunger, love of dice (gambling), and desire for women.[6]

Although Brahmā's creation story (100–300 C.E.) explains origin of the body, which is subject to actions (which are classified as good or evil), this creation of sin story shows that the gods made man (*purusha*) or the human body as a weapon against demons, and then infused him with sins, so that he might not attack them or become one of them. Sin's purpose is to prevent humans from fighting the gods or entering the world of gods. This myth reveals humans as close associates of gods as opposed to demons, but also situates them as rivals to the gods. Theologically only sin separates humans from divinity. This point is central to Hindu theodicy, which establishes the body as the purest and highest of life forms on earth. This leads us to the next issue concerning the human body: disease and death as limits intended to distinguish the human body from the divine. Therefore, I will discuss below, the divinities of disease and death, and how they influence the human body.

Creation of Goddesses of Disease and Death

Disease and death are limitations on the human body, corporeal forces that keep us distanced from divinity. As we saw above, sin most basically maintains that distinction, so the goddesses of death and disease can only affect humans in response to sinful behavior. To help understand the interaction of these divine arbiters of disease and death, with human life, I will include a short summary of the story of the creation of the goddess of death.

> Initially, Brahmā did not have a solution for the creation he brought forth. As creation increased and crowded the three worlds, he became angry and burned the universe on all sides. Then Siva met him explained to him that those that were burnt by him to ashes will never again return. And advised him that he should 'restrain his energy and devise some other means for the welfare of all creatures, so that they may return.' Then Brahmā restrained his anger. As he restrained his anger, from the fire of his energy appeared a black woman with red garments and red eyes, and Brahmā called her Mṛthyu and told her to destroy everyone.
>
> However, the goddess Mṛthyu was not happy about her duty and appealed to Brahmā that it is not fair that a woman be created for such cruel task. However, Brahmā convinced her that she would not accrue any *adharma* from her act of killing. He explained that, "creatures oppressed with disease will not blame you. The tears that you shed will be diseases, which will oppress men when their time has come. Yama, who is the eternal *dharma*, will assist you and so will the diseases." Brahmā further instructed her that, "at the time of death, you will cause the creatures to be attached to anger and desire and so you will escape *adharma*."[7]

This myth further develops on the god's struggle to keep humans away from heaven (the world of gods), resulting in Brahmā's anger, and ultimately the goddess Mṛthyu, a black woman with red garments and eyes, created from that anger for the purpose of keeping humans away from heavens. In this story Brahmā explains to the goddess grieving about her evil job that, the tears she sheds will bring about diseases, but because she perpetuates the human sins like anger and desire at death, and she would not generate any *adharma* (sin) herself. He further assured her that Yama (the god of death), the eternal *dharma*, and the various diseases would assist her. Two points are clear from the creation story of Mṛthyu: firstly, she causes death, and secondly, she causes human attachment to desire and anger, which are considered to be the most important vices that lead to sin, and thereby makes humans accrue sin and keep them from heaven.

This brings one to disease: it is what leads to death and comes from sins caused by Mṛthyu's influence on human vice. Specifically, disease is the domain of *Grāhi*, another goddess, who causes its varied forms; she, like Mṛthyu, employs worms as the tools/objects of disease, but death is not her domain. *Grāhi* causes the disease as well as cures it. We see this connection in the medical literature discussed below that treat worms with spells that appeal to *Grāhi* in medical literature.[8]

Grāhi appears as the goddess who causes dangerous diseases in both Hindu religious texts, as well as medical texts like the *Charakasaṃhita*. Her different manifestations as the *Grāhis* Revathī, Shashtī and Rohinī are said to inflict pregnant women and infants. According to the Vedic texts, *Grāhi* the goddess of disease (AV 6.112.1) ensnares her victims causing destruction (AV 1.31.2; 19.44.4; Vaj.Sam. 12.62; Śata. 7.2.1.1–17), and death (Kath. Up. 4.2.).

The prayer to Grāhi for someone afflicted with a disease contains one of the earliest references for understanding that there is a close relation between sin and disease:

> The gods wiped off their sin on Trita. Trita wiped it off on human beings. If Grāhi has reached you because of that, let the gods make it disappear. Wipe off the sins on the embryo-slayer.[9]

This verse clearly states that the gods wiped their sin on Trita, who in turn spread it on humans; thus, sin originated with gods and was passed on to humans. The above verse also states that Grāhi might affect one because of "that," the sin that was passed on. It also indicates sin could be transferred to another Grāhi goddess known as Shashti "the embryo-slayer." All "sin" comes from the gods (see the creation of sin story above) and, although the goddess of death (Mṛthyu) only inflicts disease on a sinner, while the goddesses of disease (Grāhi) can either afflict people or cure them by wiping off their sins. Both of these goddesses, Mṛthyu and Grāhi, act as the commanding forces in sending disease, and removing the disease respectively, and the worms seem to be tools/objects for this purpose. As seen in the creation story of Mṛthyu, diseases originate from the tears she shed upon her hearing about her duty of killing the people. Grāhi as noted above causes disease to clear away people's sin, since she wipes off their sin in the process. Therefore, both these goddesses cause diseases (manifest as worms), but with different goals. However, none of their activities would exist without the first cause of sin, which humans (*purusha*) inherently contain in their bodies (*śarīra*) and actions (*karma*).

Although the gods made sin, people can still work in this world to fulfill their *dharma* and eventually attain heaven through a long process of purification. One type of purification involves removing sins by undergoing the suffering from diseases, a condition attributed to worms and under the purview of these two goddesses, Mṛthyu and *Grāhi*. The creation of worms, goddesses of disease and death, and sin according to early Hindu texts and their relationship to the humans and cyclical life, imply that early India concepts of disease were primarily religious.

Summation: Sin and Disease

In this section, the Hindu myths discussed specifically indicate that disease in Hinduism arises from sin and manifests in earthly life, instead of resulting from material based causes. This religious conception of disease relies on the view of the body as simply the vessel manifesting the result of actions in past lives. It separates the body from its current actions, focusing instead on the effects of past-life actions of the incorporeal subtle body on its present earthly shell. This understanding gradually developed in Hinduism. Early Vedic religion held a positive view of life aspiring for heaven, invoking benevolent gods, with no cosmic mechanism to stop the mass of humanity from overcrowding the heavens.[10] Later Vedic literature, especially the *Upanishads*, included descriptions that portray a sinful nature of the body that generates bad *karma*, as implied in the creation of sin and Mṛthyu myths above. The rest of Hinduism, as depicted in the Epic and *purāṇa* literature, builds on this convention of *karma* affecting the body, while correlating it with the positive goals of Early Vedic religion. The resulting prevalent view in Hinduism states that the negative beings, such as worms, exiting in this world can evolve into more positive forms, like humans, and enjoy a better life through practice of *dharma* (successfully fulfilling one's predestined cosmic role or assigned work), or through performing rituals.

Previously, I showed how the body and worms were created and the conditions that allow worms to invade the body, and which deities have command over disease and death. The next question is then how a diseased body is treated and what kind of socio–religious and administrative roles are permissible under such condition. Therefore, next we examine the practices of these religious rules with regard to worms and the diseased body in India.

Worms and the Diseased Body on Earth: Practice

In the following section, I will explain how the myth based attitudes discussed above, towards worms in particular and disease in general, influenced religious, administrative, legal and social practices of India. This will proceed in two sections: the first discusses the social, religious and political treatment of diseased body/person, while the second examines the hierarchy of worms/disease causing organisms.

Wormy Corpus in Social and Religious Life

Whether it becomes infected or not, Hindu teachings describe the human body as under constant attack and subject to torment by tiny worms as early as its time in the womb. *Garudapurāṇa* mentions that worms constantly bite the fetus in the womb (Ga.pu. VI.9–10). This explains the nature of the worms that are born to torment or afflict in accordance with the severity of the sins of humans. Since in life a person may come across great varieties of worms, these torments and troubles tend to increase as one ages and is subjected to other factors of disease and death. A number of sins are listed in the *purāṇas* for which one is born as a diseased person: slayers of women and destroyer of embryos (Ga.pu. 5.4.) and those who eats the flesh of other beings becomes savages full of diseases (Ga.pu. 5.10).

However negative the association, a diseased body does not always mean an automatic dismissal from work, family or community life. An interesting fact noted in the *Arthaśāstra* indicates that lepers can continue to work for the army. Indeed, one section of the *Arthaśāstra* is devoted to the preventive and curative medical care of the army, and includes a potion formula that is said to cure leprosy (Arth. Book XIV. IV. 425); this implies that a diseased body did not mean disqualification from serving in the army. Lepers in this early period appear not to be treated as outcasts as they were later, but were allowed to serve the state and were provided with medical care. A Vedic scholar with a diseased body was not supposed to be shunned, but treated on par with any other Vedic scholar and be given donations of food when requested (Manu 11.1); even though the diseased body was seen as impure, the power of Vedic scholarship overrode this negative status.[11]

Manu even includes an ill man (diseased body) in the list of people deserving respect (Manu 2. 138). Manu also suggests that a householder may feed an ill person who arrives at his house as guest (Manu 3.114) without hesitation. Marriage of a leper (woman) is also allowed, and her father is excluded from paying any fine if he announced the flaw before marriage; if he did not announce the flaw before marriage, the fine would be one hundred *nishkas*[12] (Manu. 8. 235), but the marriage would not be dissolved and cannot be considered illegal. Leprosy is a disease said to affect the body in visible ways and prevent the diseased body's participation in normal life. However, the injunctions noted above in one of the widely used books of India, indicates that the practice with regard to a diseased body is that of tolerance. Forgiveness is one concession allowed in favor of a diseased person; Manu enjoins that a well wishing king should forgive an ill person even though the person was to insult the ruler (Manu. 8.312). This shows the society's tolerant attitude toward those who possess a diseased body and functionally compromised. Diseased persons, however, are not eligible to become witnesses (Manu 8. 654, 77) and are excluded from certain religious rituals such as the *Śrāddha* (death rituals): if a leper looks at the priests while they are eating at a ritual, the donor loses the reward for feeding a hundred Vedic priests (Manu. 3. 177). Therefore, the leper is not allowed in the vicinity of where the *Śrāddha* ceremony is being conducted.

Medical Treatment for the Diseased Body

A diseased body is permitted to reside amongst one's own family and community, even allowed to perform certain religious rites and work for the military partially because of the potential for active medical treatment. Medical diagnosis and treatments of bodily infections are noted since 1000 B.C.E. in India, however evidence regarding establishment of medical centers in cities dates from the Maurya Empire (325 B.C.E. onwards). From that time, we find in the *Arthaśāstra* and Aśoka's inscriptions mention of the establishment of hospitals for the treatment of humans as well as animals and planting of medicinal trees. Medical texts include treatment of a variety of diseases, including those caused by worms (*krimi*), virus (*visuchika*) and bacteria (*masuchika/pox*), while also discussing the use of amulets and magical formulas to appease the Grāhi goddesses.[13] Suśruta also includes leprosy and worm disease (*krimigadha*) among the numerous other diseases that could be cured by consuming garlic and turmeric, noted antibacterial and anti-helminthic substances.[14] This inclusion of both medical and religious formulae raises the question of the relationship between the material and religious worlds, worlds both inhabited by worms.

Looking more specifically at the convergence of religious and medical therapies, we find variety among the various sources. In the sections of the classical texts covering worms (*krimi*) appear medical procedures relating to diagnosis and treatment with herbal medicine, but the sections on fetal and childhood afflictions of Grāhis also deals with diagnosis, but instead provides religious formulas for charms and amulets as cures.[15] This indicates a dual level treatment system. The text advises using medicine to address the disease causing organisms (worms etc.), while simultaneously attending to true religious origin of the disease (the *Grāhi* goddesses) with utterance of magical formulae and amulets. In this case, we see a distinction between the treatment of the physical side of disease that requires herbal treatment (worms) and the religious side (*Grāhi*) that requires magic.[16]

The practice of treating a diseased body simultaneously for wormy afflictions and *Grāhi* afflictions show a full confluence in the religious and medical practice with regard to disease. It also indicates that the disease is not only a physical manifestation, but connects to cosmological realms. This overlap also relates back to the correlation of sin and worms previously discussed. Since *Grāhi* afflicts and seems to have command over sin (by cleaning it or sending worms to clean it), then praying to her might help one get better, like in the spell quoted in the second main section where Trita wiped off his sin on *Grāhi.* People thus pray to her in the hope that she will wipe their sins off and clear their diseases.

Aside from overlap in therapy, Kautilya's administrative text *Arthaśāstra* (500 B.C.E.–300 C.E), one of the foremost and earliest administrative treatises of the world, exposes the social and spatial intersection of medicine and religion. Kautilya discusses the location of medical complexes, or the hospitals, constructed in the northwest of the city near and to the right side of the residences of Brāhmans and royal palace complex (Arth. Book II. IV.55). This strategic position of the hospitals reveals the similar importance and connection of both the political/administrative and medical establishments in ancient India. This position is also important according to the religious conventions. It can be further noted that the medical center is located in the north, a direction the Vedic texts indicate as beneficial for sick people to travel.[17] It is logical to integrate the two disciplines of medical and religious spheres, for the disease conception of that period not only connects with the physical world of the earth, but also the cosmology through the intervention of the disease causing goddesses.

Arthaśāstra also describes the temples of divine Physicians (Aśvins) and Vaiśravaṇa (Kubera, lord of the Yakṣas) located at the city center (Arth. Book II. IV. 56) and of Sri Madirāgraham (goddess of Madirā Revathīgraha or Revathi wife of Balarāma is a noted *Grāhi* afflictor of children and pregnant women) (Vemsani, 122–123). These two centers (medical centers in the north/northeast with the Brāhmans and temples of magical and afflicter deities in the center of the city) reflect the two types of concurrent religious and medical practices then prevalent in dealing with the diseased body: medical treatment by scholars (also trained in ritual) and divine intervention. This again shows the relation of disease to sin and its cure lying in appeasing the goddesses, while also treating the worms.

Worms: Disease Causing Organisms

As explained above, prevalent notions in Hinduism suggest that because of past sins one can be reborn as a worm or afflicted with a wormy corpus. Here I will discuss a number of sins leading to rebirth as a worm or to infection by worms. These punishments for sins operate within the Hindu system of *karma* and reincarnation, where sins of varying severity and kind result in a range of possible rebirths, with one of lowest being as a worm.

We find a number of sins resulting in rebirth as a worm in the *Laws of Manu*. Stealing property causes a man or woman to be reborn as a worm in the excreta for sixty-four hundred years (Manu.12.14), with the women, of course, becoming female worms (Manu.12.69). Manu holds educated priests to an even higher standard indicating the severe punishment for a priest drinking liquor entails entering the womb of a worm, bug, or moth, or of a bird that eats excrement, and of violent creatures for rebirth. (Manu.12.56)

The particular association of stealing with rebirth as worms also appears as a widespread phenomenon demonstrated by its inclusion in social contracts of numerous official inscriptions and donation plates from temples. This connection accords with prevalent customary laws noted in the *Manudharmaśāstra* and the *purāṇa*s. *Garudapurāṇa* mentions that a person "who takes away a plot of land, which was given away by himself or by another, is reborn as a worm living in the excreta for sixty thousand years" (Ga.pu. V.45). The inscriptions and donation plates end with a curse that promises infestation by worms, or other similar punishments, if the terms of the inscription or donation are violated. Such inscriptions are clearly part and parcel of the classical legal tradition, for those from 300 CE repeat the exact words of the verse from *Garudapurāṇa*, 'Who takes away the land, given by self or another, lives as a worm for sixty thousand years in excreta.'[18] This use of worms as a grave threat found in many locations, and especially associated with religious institutions, reveals the common and powerful symbolism of the lowly worm. It appears as an exemplar of sin, with such a curse generating great disgust and fear. This common curse indicates the purgatory nature of rebirth as a worm and how worms have similar roles the religious and administrative spheres of the society. Stealing is seen as one of the worst crimes. Especially despicable is stealing donations given for the welfare of society or even in the name of God. Stealing such money is tantamount to stopping the spread of happiness, and therefore a very negative crime. The punishment is not related to the actual value of the theft or damage, but connected to the intent of causing damage to the greater good of the community resulting in this type of wormy birth. This suggests that violating the social contract results in ceding one's privilege of being a human, resulting in rebirth as a worm.

However low the worms may be, one is not allowed to harm worms unnecessarily. If one harmed worms unintentionally, he/she should perform rituals of purification immediately, not to accumulate bad *karma*. Even so, the ritual prescriptions reveal a lowly regard for worms. If a person injures boneless beings (such as worms), he regains purity by simply performing *prānayāma*, a controlled breathing practice (Manu. 11. 142). In contrast, Manu require a greater penance for the minor action of cutting fruit trees, shrubs, creepers, lianas, or flowering plants, declaring one hundred *ricas*, or double verses, must be uttered (Manu 11.143).[19]

This difference raises the question of the true cosmic rank of the worm. *Manudharmaśāstra* lists three levels of existence (Manu. 12.41) for life on earth, and states that "stationary objects, worms, and bugs, fish, snakes, turtles, livestock, and wild animals are the lowest level of existence to which darkness leads" (Manu.12.42). The *Sangam* literature classifies living beings on the earth into five types based on their senses: the last is trees with one sense, and worms are classified as second type with two senses (touch and taste), while humans are classified as the fifth and highest class of beings, possessing five senses.[20] Even in light of these structures, the above punishments reveal worms as especially lowly, because the value of their life is lower than that of cosmically simpler creatures. From this perspective, worms are the lowest of living beings, lower than immobile plants, and thus the complete opposite of the highest of the living beings on earth—humans.

Summation: Worms and Purity

Based on my examination of the religious, social and medical texts, I have discussed the social, religious, medical and legal practices in this section. Figure 1 summarizes these practices concerning the worms and wormy corpus in India in general and Hinduism in particular.

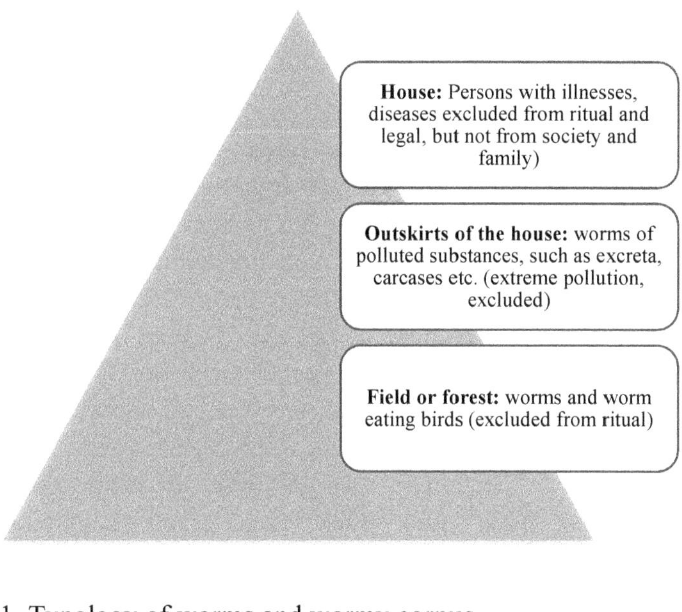

Figure 1. Typology of worms and wormy corpus

The various levels of impurity and exclusion form a neat pyramid with the field or forest as the largest and least pure zone, outskirts or outside of the house forming the middle zone of impurity with some area of purity and sanctity, while the smallest zone, the home, has the highest restrictions on impurities. This pyramid clearly demonstrates that the level of purity correlates to the presence or absence of worms; worms therefore are not only a marker of disease, but also a reflection of social and religious concepts of purity and exclusion. The house is the purest zone of all, since it is clean and may not contain any worms. A diseased person residing in the house is of course not pure and excluded from religious rituals, such as *Śrāddha*. The forest contains numerous worms and wild animals and is the most impure zone. Forest beings and Birds are excluded from any sacrificial offering and are considered inferior food, because they subsist on a diet of worms, which may in turn subsist on excreta or other lesser food and undesirable food. Vedic rituals also prohibit sacrificing forest animals.[21] The next level of prohibition is against diseased persons or persons dealing with death and polluted substances (excreta etc.), who are therefore not allowed participation in any rituals. Due to their perceived proximity to the impure they are outcast for religious purposes. To be born as a human being is still a privilege when compared to forest creatures, or worms, hence a person is not completely excluded from family or social life: he or she may marry, donate, or do other social activities, in order to gain merit and a better rebirth.

Although a person with a diseased body is considered physically impure, a state also indicating spiritual corruption as a result of past actions, it is still proper for them to seek medical help as part of their process of purification. This is also the case because the four goals of life include *kāma* (pleasure) and *artha* (prosperity), for even the impure should strive for enjoyment rather than just passively enduring their allocated suffering. The rules of *karma* act as a guide for how to live a positive life with minimal damage and harm to creation, and the punishments of this just order of *karma* compensates for the absence of the devil or a world of evil in Hinduism, explaining the existence of both worms and those who suffer with their infection. This functioning of sins and worms explicates the interconnection between the three worlds of heaven, hell and earth, the topic of the following section.

Worms, and the Corporal Body in Cosmology: The Connections

I discussed in the above two sections the literary evidence concerning the thought (origins: body, worms, sin, goddesses of disease and death) and practice (worldly realm: rules, medicine and rituals) with regard to worms and the body. In this section, I will explore the linkages within the cosmology where bad *karma*/sin (sometimes manifesting as worms or a diseased body) acts as the variable that connects the worms and the body from one world to the next. Therefore the following is a discussion on the interrelationship of these three, disease (worms), deities (other worlds) and sin (as a link between this world and other worlds).

Worms and Sin in the Worlds (Saṃsāra)

Worms play an important role in this cyclical nature of *karma*, and worms are found in hells, in addition to on the earth, eating away the past sins, since the soul can manifest in any form based on its *karma.*[22] In the cyclical, transmigratory nature of life in Hinduism the key component is the *atma* (soul), which remains constant, and unchangeable (*Bhagavadgīta*.II.20). Because of the nature of this soul, sin in the form of *karma* can pass from one world to another.[23] Therefore, the soul can be punished for sins according to the system of *karma* on earth and in other temporary worlds like hell. According to this cosmology, the end of life on earth is only the end of the mortal body, but not the end of the soul itself. However, even predating this vision of *karma* we find otherworldly punishment that anticipates the roles of worms and consumption in cosmic punishment. I will include below the thought of Hindu cosmology to understand the development of these cosmologies and the unique role that the worms undertake.

I will discuss below the nature of other worlds, in order to understand the evolution of the transmigratory nature of sins and soul. *Karma* is a result of the actions (good or bad), including invisible sins or merits, in the past lives that visibly manifest on the earth, as well as other worlds like heavens and hells (Ga.pu.5.51–56). *Karma* is a unique feature of Indian religions and plays a major role in the Hindu, Jain and Buddhist theology; it excludes the need for a concept of substantive evil to explain suffering. Scholars have noted that, "the most complete formal solution to the problem of theodicy is the special achievement of the Indian doctrine of *karma*, the so-called belief in the transmigration of soul,"[24] and "for Hindu thought there is no problem of evil,"[25] as *karma* eliminated the need for evil.

Excluding the substantive evil does not exclude other worlds for punishment, like the different temporary hells. Hells in this system function to reform sinners before they are reborn on the earth, repaying their sins while accruing positive *karma*. Still, as suffering in the hells does not completely destroy a person's sins, the sinners reborn on earth must further endure the *karmic* retributions of previous actions and may face disease or rebirth as miserable creatures like worms. Whatever the manifested form on earth may be, either as worms or as a human, the soul is the same and the *karma* continues to follow one on earth. It is the soul that completes this circle of life, as noted in the *Bhagavadgīta.*[26] In order to fulfill this circularity of *karma,* Hinduism includes numerous hells and heavens, along with earth in the universe. The cycle of births and lives in these different worlds is unending, unless one's *karma* is completely annihilated and one obtains the liberation of *moksa*.

The Development of Hell

Early Vedic literature does not include a developed view of the different worlds and the word "*loka*" is used loosely to indicate, a place, residence or a villa or grove. Still, even without the cosmology of the *karmically* driven movement of the soul between worlds, existence in multiple worlds appears in these early myths.[27] The travels of Brigu and Naciketa to the other worlds described in the Vedic literature (JB 1:42–44) demonstrate the early Vedic understanding of connectivity of life through different worlds.[28] The story, "Brigu's journeys to the other worlds," proffers some of the earliest evidence concerning the cosmological thought of Hinduism, connecting earth to the heavens and hells.[29] This story has various interpretations, but it is clear that it incorporates the early views of balanced retribution. These other worlds reveal a reversal of roles seen on earth: those that were eaten here on earth become the eaters in the other world; for those who kill trees in the afterlife will be eaten by trees; those who eat animals will be eaten by animals; those who cook grain will be eaten by grain, etc. In later writings, worms appear in a similar capacity as iconic eaters, when the *purāṇas* colorfully describe them as the manifestation of the sinful or the food of the sinful in hells.

In the fifth of the six worlds included in "Brigu's travels," we find a dark man guarding two streams of blood and butter. Though not an eater, he punishes the killers of men and represents an early depiction of Yama (the god of death, in charge of dispensing punishments or rewards based on one's earthly actions in the later *Upanishads*). In the fourth world, the two women, identified as Faith and Non-faith, guard a treasure and are none other than the goddesses of disease and death—Grāhi and Mṛthyu.

This early conception, with Yama and the goddesses only punishing certain crimes, evolves along with the development of *karma*-based cosmologies. The famous Upanishadic story of Naciketa's travels to the world of Yama (KU 1–6; TB 3.11.8.1–6) appropriately demonstrates the important new role Yama acquired in this cosmology, as he become the dominant figure representing the cycle of life for beings on earth. In this story, we see he is the arbiter of the fates of sinners in the heavens and hells, and thus directs all movement of souls between earth, heaven, and hell. Even though he sends people's souls to hell, in the cyclical system of *karma* one's stay in Yama's hells is only temporary.

In the hellish worlds of Yama, worms also come to have a place. This logically results from worms' status in all the ancient India religions (Hinduism, Buddhism and Jainism) as the basest creature in the cosmic hierarchy.[30] They lie opposed to humans, who as the highest living creatures are just a single step below the gods (Vi.pu. 2.6.211; Ga.pu. 6.40). Worm's impurity and lowliness personifies the very sin that keeps humans from the divine (Manu 12.14; 12.16). Yet, although worms signify this negative state of pollution, they are not evil. Hinduism does not have a concept of evil comparable to the Western religious traditions. There is no world of devil or demons, but there is only a world of gods. The hells under the leadership of Yama reside in a divine cosmology, and not a demonic cosmology. The closest equivalent of evil in this world is '*pāpa*' (sin), which resembles a temporary taint. Worms play a major role as perpetrators of the taint negative *karma*, and though not demonic, participate in the punishments of those in hell.

Hell in Hinduism (Vi.pu.2.6; Ga.pu. 6.60–62) resembles a correctional facility headed by the Yama, known as Dharmarāja (king of duty), and though temporary it remains a horrific world of terror. Anybody that trespasses dharma and accumulates sin is subject to the punishments meted out by Yama Dharmarāja (almost like a judge), but as mentioned above, these punishments in hell cannot fully purge the sins, so the soul is reborn on earth for a second chance to cleanse the sin and even accumulate the merit to move up in the cosmic hierarchy.

Worms and Hell

As the worst rebirth among eighty-four hundred-thousand (8,400,000) possible rebirths for the sinful *Garudapurāna* lists, the existence of the worms exemplifies sin's presence in this world. This is true in both earth and hell. The third main section demonstrated how by committing certain sins a person is born as worm in on earth, now we turn to the types of sins leading one to be become a worm or worm eating being in the hells. *Krimibhakṣa* (hell of eating worms) hell is where worms are the food of the sinner; this hell is the destiny of those who hate parents, Brahmans, gods, and spoil precious gems (Vi.pu. 2.6; Br.pu. 205). *Krimīsa* (hell of worms) is for sinners condemned to live as worm. One receives this punishment for practicing magical rites to harm others (Vi.pu. 2.6; Br.pu. 205). *Sūkara* (swine hell) is where the sinner is destined to live as a pig consuming refuse and waste that might contain various worms or germs. One finds punishment in this hell for killing a Brahman, stealing gold, or drinking wine (Vi.pu. 2.6; Br.pu.205). However as mentioned above, the negative *karma* of the sinners is not expiated here by this hellish life. They are again born on the earth to undergo their suffering, while also possibly gaining positive *karma* to obtain heavens or better (unite with Brahmā or obtain *moksa* (liberation)). Like with rebirth as a worm on earth, sinners punished with wormy lives in hell still can potentially purge all their sins to become the true opposite of the worm—divine.

Summation: Worms, Worlds, and Sin

The only quality that separates humans from that of deities is sin, which according to mythology was created by the gods (see the second section) and infused into man (purusha) to exclude him from heaven. Therefore, like three intersecting circles (see figure 2), birth, disease and death decide the path of one's life in this world. The eternal cycle of life and death (*samsara*) is intersected by the circle of disease or worms (representing the sins). Sin and its expressive forms of disease or worms follow one in all the hells and earth.

The Vedic myths first established the interrelatedness of the universe (divine and human worlds), the thought of the sins, and the idea of repaying for the sins. In the later myths this relationship is not only maintained, but also greatly elaborated. The roles of Yama Dharmarāja, (the eternal *dharma* who maintains the world of death and punishment for sins) is expanded, as is that of the two other goddesses, Grāhi and Mṛthyu, who ensnare beings and mete out punishments for past sins. In this process the role of the worms also enlarges. They come to represent the life of sinners either as worms or as diseased body, in the hells or on earth as shown in Fig.1.

In this cosmology worms come to symbolically encompass all types of sin: the wormy diseased body is emblematic of *karmic* suffering in human life, while the disease-causing worms represent the worst possible rebirth. Both sides of this system of punishment originate in sin: the worm that infects and the diseased body are simultaneously paying for past sins, although the gravity and type of sins may differ in each case. The cause of the disease, i.e. worms, and the diseased state of the person, are both results of sin, but more positively their mutual suffering is a path to purification (Ga.pu.6.60–64).

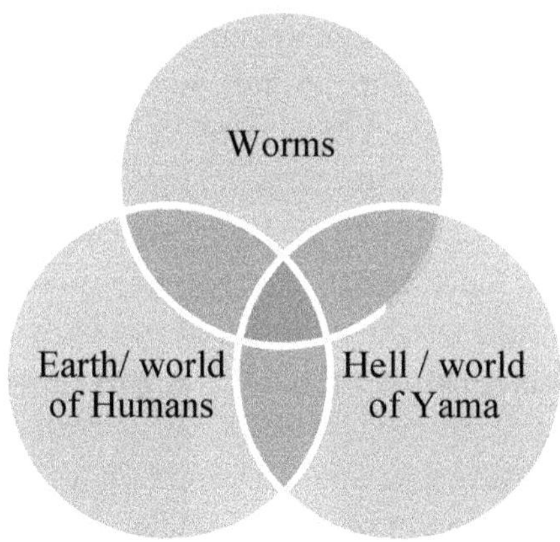

Figure 2: Circle of Worms and the Body

Conclusion

The myths and social practices concerning worms studied in this paper establish an important connection between four factors: the interrelated role of worms and the body; the confluence of worms and sin; the association of socio-religious and administrative structures with the diseased body; and the place of worms in a universe with transmigration. The individual body is not just subject to physical rules but also cosmological and theological rules dictated by Hindu culture. Therefore, individuals in India are not just individuals, as McKim and Mariott (1968) have persuasively shown, for their body relates to more than one phenomenon simultaneously. They relate to three levels: personal (individual), social (transactional) and religious (universal self). The human body then includes the social being and also the divine being. The divine individual aspect exists in the body distorted by the presence of sin and the combination of these two components dictates and carries out the actions of this life, connecting the body to past and future lives. Worms are crucial symbols of the negative aspects of the world, or sins, and play an important but detested role in this multilayered order of life. Numerous rules of purification (ritual washing and oblation rituals) show this fear of their corrupting force. However, worms and sin are also an indispensable part of creation, as seen from the creation myths. Worms are the lowest of the creatures with impure work such as infection and disease, and contrast with the human body, the purest and highest living being.

The post-Vedic socio-religious concepts are led by the common *saṃsāra/mokṣa* dichotomy. The sin indicates the processes that lead to and sustain these states. A diseased person, by default is a defiled sinner, and is excluded from taking part in the *Śrāddha* and other rituals, prohibited from giving witness, or learning the *Vedā*. He is excluded from administrative and religious functions of the society that require unimpaired thought and action. However, sin or disease (in other words, attack by worms) is inevitable and Hindu attitudes dictate that one should live the best possible life and purify one's sins through passing through sickness. Therefore, a diseased person is not excluded from having a personal life as we have already discussed; he or she is allowed to marry and to seek healing. Disease as a marker of sin stands as an important socio-religious signifier, but it does not make one irredeemable.

Given this dimension of worms and creation, attitudes and behavior towards worms and disease can indicate a person's relationship to the created Universe. While the human body is distinct from the divinity only because of the possession of sin, the worm is the lowest and represents sin on earth as well as in hell. Humans are duty bound to live as purely as possible to minimize the risk of accruing sins to avoid being born or infected by worms. Worms even though lowly are their own individual entities, with social, cultural and religious roles attached to them, closely tied to humans and sin. This study concludes that worms play a central role in Hinduism as the epitome of sin its relationship to the body and cosmology.

Notes

1. Although Buddhism and Jainism do not include a belief in the existence of soul, circularity of life and rebirth are important aspects of these religions. Jainism includes an elaborate and hierarchical classification of matter (*pudgala*), heavens, and hells, which constitutes the universe. The legend of Mahavira and the Buddha include numerous past lives and travels between different worlds. For more information see, Carl B. Becker, *Breaking the Circle: Death and the Afterlife in Buddhist Thought* (Carbondale, IL: Southern Illinois Press, 1994); Paul Dundas, *The Jains* (New York: Routledge, 1992); J.E. Cort, *Jains in the World* (New York, Oxford: Oxford University Press, 1998), 116.
2. Besides worms (helminthes), four other kinds of microbes could also be included in this consideration: bacteria, viruses, protozoa and fungi. From a human perspective, some of these worms work for the better, some for the worse and some seem to simply take up space that might otherwise be occupied by something harmful. *Bibliodata* defines a worm literally as any of numerous relatively small, elongated soft bodied animals especially of the phyla Annelida, Chaetognatha, Nematoda, Nemertea, Platyhelminthes, and also many insect larvae, and goes on to include, " a person who has a nasty or unethical character undeserving of respect." "Worm," accessed, July 4, 2011, http://www.bibliodata.com/biblioSearch/?term=worm.
3. Primary sources include: *Atharvavedā* with the commentary of Sayaṇa (Geetha Press: Bombay, 1895); *Arthaśāstra of Kautilya*, trans. R. Shama Sastry (Mysore: Mysore Printing and Publishing House, 1967); *Bhagavadgīta*, trans. Edwin Arnold (London: 1885); "Katha Upaniṣad" in *One Hundred and Eight Upanishads* (Bombay: Geetha Press, 1913); *Mahabharata*; *Maitreyani Samhita*, ed. L. Von Schroeder (Wiesbaden, 1881); *Manudharmaśāstra: Sacred Books of the East*, trans. George Buhler (London: Oxford University Press, 1886); *Rigveda, with Commentary of Sayana*, trans. Ralph T. H. Griffith. (London, 1890–92); *Śatapatha Brāhmaṇa: Sacred Books of the East Parts 1–5*, trans. Julius Eggeling (London: Oxford University Press, 1882–1900); *Taittiriya Samhita*, with commentary of Sayana (Calcutta: Royal Asiatic Society, 1859); *Taittiriya Brāhmaṇa*. with commentary of Sayana (Calcutta: Royal Asiatic Society, 1859); *Vājasaneyi Samhita, The Upanishads, Part One: Sacred Books of the East*, trans. Max Mueller (London: Oxford University Press, 1952)
4. *Saṃsāra* is the brief passage of soul through life on earth. The earthly actions of the living beings determine the course of next life for the better or worse. At no time is life on earth entirely separated from the cosmic circle. For further information see, K.H. Potter, *Presuppositions of India's Philosophies* (New Delhi: Prentice Hall,1965); Swami Satprakashananda, *Methods of Knowledge* (London: George Allen and Uwin Ltd, 1965);

R.S. Khare, "Annambrahman: Cultural Models, Meanings, and Aesthetics of Hindu Food" in *The Eternal Food: Gastronomic Ideas and Experiences of Hindus and Buddhists,* ed. R.S. Khare (Albany, NY: State University of New York Press, 1992), 201–221

5. Anger and desire are seen as very dangerous sins since they lead to numerous *vyasanas* (vices) that result in negative *karma. Manudharmaśāstra* lists 10 *vyasanas* arising from desire (VII.47) and 8 from anger (VII.48). In another list in *Manudharmaśāstra* four *vyasanas* spring from desire (VII.51) and three from anger (VII.52). This second list also appears in *Arthaśāstra* (VIII.3) even though the vices are listed in a different order.
6. JB.1.97–98
7. Mbh. 12.248.13.21; Mbh. 249.1–16
8. *Grāhi* is not a single goddess but a category of goddesses that could catch people unawares. The Sanskrit term *Grāhi* or *Grāha* translates as the "grabber." Revathi and Shashti are well known *Grāhi* goddesses, while there are numerous others not well accounted or noted. Numerous other goddesses of disease are also known from popular culture in India, including Śītalā (small pox goddess), Mariamma (small pox goddess literally 'the cool one'). For detailed study of these goddesses, see M. Moreno and M. Marriott, "Humoral Transactions in two Tamil Cults: Murukan and Mariamman," *Contributions to Indian Sociology* 23. 1(1989): 149–169; Edward C. Dimock, "A Theology of the Repulsive: The Myth of the Goddess Śītalā," in *Divine Consort* (Berkeley, California: Graduate Theological Union, 1982), 184–209. Pauline Kolenda, "Pox and the Terror of Childlessness: Images and Ideas of the Smallpox Goddess in a North Indian village," in *Mother Worship* (Chapel Hill: Univ. of North Carolina Press, 1982), 227–250. Following this tradition of associating goddesses with disease, the worship of goddess Aidsamma began on November 30 (World Aids Day, 1998), in a small temple in the village in Menasikyathannahalli, in the state of Karnataka, since Aids is now becoming more common than that of smallpox. *The Hindu* accessed: July 12, 2011: http://www.hindu.com/2007/12/09/stories/2007120957950500.htm.
9. *Tait.Br.* 3.2.8.9–12; *Mait. Sam* 4.1.9; AV 6.113
10. Gonda, 25–41and 130–43
11. This does not mean that a leper is allowed to enter a Vedic school and learn the Vedas. It only means that the Vedic scholar was afflicted with leprosy after the completion of his learning, and still commands respect for his learning. The leper should be excluded from learning the *Vedā* (Manu. 3. 161).
12. *Nishka* is copper coin in Ancient India known from 1000 B.C.E onwards. It is valued at 1/26 of the value Silver coin, the *Karsapana* and 1/96 of the gold coin *Suvarna.*
13. Wuzastyk, 136–139,160, 286
14. *Ibid.* 154–160
15. Filliozat (1964)
16. Medical treatises of Srilanka also follow similar precedents in treatment for both wormy infections and *Grāhi* afflictions, although the specific religious conventions of Srilanka are unique. See, J. Liyanaratne, "Sinhalese Medical Manuscripts in Paris," in *Medical*

Literature from India, Srilanka and Tibet, ed. J. G. Meulenbeld (Leiden: E.J. Brill, 1991), 73–91.

17. Charles Malamoud, *Cooking the World:Ritual and Thought in Ancient India* (Oxford: Oxford University Press, 1996), 79.
18. Translation is mine. *Svadattaṃ paradattaṃ va Yo harethi vasundharṃ shashṭiṃ Varshaṃ sahasrāṇi vishtāyām Jāyate krimiḥ.* V.V. Krishna Sastry, *Select Epigraphs of Andhra Pradesh* (Andhra Pradesh State Department of Archaeology and Museums: Hyderabad, 1985), 32–33
19. Even though worms are ranked higher than trees and shrubs in the hierarchy of the living beings as they can move, in case of any harm done to the worms, the punishment is trivial, even less than that of the punishment in the case of harm done to beings ranked below them, the trees. This shows the contempt for worms arising from the view of them as useless and even harmful. This difference also relates to the relative sacredness of these beings. Even though worms can move, and are thus more advance, trees are more sacred following the conventions of the Vedās. In the Vedā texts the goddes Vāc (sacred speech) left the gods and hid in waters and then in the trees (*Tandya* 6.5.10–13; *Tait.Sam.*6.1.4.1; *Mait.Sam.* 3.6.8; *Kath. Sam.* 23.4), and as a result water and trees became more sacred than other earthly living and nonliving things. As sacred objects injuring fruit trees, shrubs, creepers, and flowering plants merits a more severe punishment than for that of killing worms. Worms are the antithesis of the sacred and the lowest level of an incarnated soul.
20. A.K. Ramanujan (1985), 239
21. Malamoud, 80
22. For more details on *karma* and its application to life and Hindu theology see, Christopher Chapple, *Karma and Creativity* (Albany, NY: State University of New York, 1986).
23. The individual soul or self, referred to as *atma*, is made up of three Sanskrit syllables variously derived from '*an* (to breath), at (to move), and vā (to blow). *Atma* is also refers to the 'breath of gods,' in the *Rigvedā* (RV.5.168.4), and also in 'where was the breath (*asu*) the blood (*asṛg*), the essence (*ātma*) of the world?' (RV.2. 64.4). Monier Williams, *Sanskrit-English Dictionary* (Oxford University Press: Oxford, 2008 (revision), 135.
24. Weber, 145
25. Watts, 35
26. For the soul there is neither birth nor death. Nor having once been, does he ever cease to be. He is unborn, eternal, ever-existing, undying and primeval. He is not slain when the body is slain. The soul is unborn, eternal, ever-existing, and immutable. As a person puts on new garments, giving up old ones, similarly the soul accepts new material bodies, giving up the old and useless ones. The soul can never be cut into pieces by any weapon, nor can he be burned by fire, nor moistened by water, nor withered by wind. The individual soul is unbreakable and insoluble, immutable and unchangeable. Knowing this, you should not grieve for the body. (*Bhagavadgīta* II. 22–25).

27. Jan Gonda, *Loka: World and Heaven in the Veda*, (V. Noord-Hollandsche Uitgevers Maatschappij: Amsterdam, 1966), 25–41. World as a title is used almost synonymously to refer to a class of beings, such as world of animals (*pasuloka*), *Indraloka* (world of Indra), kingdom of Indra, almost similar to the modern use of world in the names of amusement parks such as Disney World or Sea World. *Loka* denotes a tranquil place in the Vedas.
28. Bodewitz, "Brigu's Journey in the Other Wold," *Jaiminiya Brāhmaṇa* I. 1–65, Translation and Commentary, With a Study of the *Agnihotra* and *Prāṇāgnihotra* (Leiden: Brill, 1973), 99–109; Lommel "Brigu im Jenseits." Paideuma, 4 (1950), 93–109; Max Weber, "Eine Legende des *Śatapatha* Brāhmaṇa uber die Strafende Vergultung nach dem Tode," *Zeitscrhrift der Deutschen Mongenlandischen Gesellschaft* (1855), 237–289; O'Flaharty, *Tales of Sex and Violence* (Delhi: Motilal Banarsidas, 1987), 33–35. She interprets his travels as going in the four directions and the up and down, based on *Śatapatha Brāhmaṇa* version, which mention the sides instead of worlds.
29. *Manu* lists 9 hells, while *Mārkandeyapurāṇa* lists 21 and *Bhagavatapurāṇa* lists 28.
30. Through good actions (merit/ good *karma*) the human soul can attain release and join the Brahmā and enjoy the pure everlasting bliss (*satcidanada*), which is a status higher than the gods. Although Buddhism and Jainism do not agree with this theory of *Mokṣa*, the Buddhist *Nirvāṇa* and Jain *Kaivalya* implicate similar connotations. Through right life and meditation, a human being can attain the greatest state of knowledge and bliss. W.N. Brown, *Man in the Universe: Some Continuities in Indian Thought* (Berkeley: University of California Press, 1966); P.S. Jaini, *The Jaina Path of Purification* (Berkeley: University of California Press, 1979); D. Kalupahanna, *A History of Buddhist Philosophy* (Honolulu: University of Hawaii Press, 1992).

The Wasp and the Worm: Getting Inside the Body of the Sage

Mark G. Pitner

> What lessons can we not learn from her transcendental chemistry! With an imperceptible drop of her poison fluid, she straightway renders her prey incorruptible! Incorruptible, did I say? It is much more than that! The game is brought to a condition which prevents desiccation, leaves the joints supple, keeps all the organs, both internal and external, in their pristine freshness and, in short, places the sacrificed insect in a state that differs from life only by its corpse-like immobility.[1]

The founder of modern Western entomology, Henri Fabre, known for his passionate research as well as a compelling writing style that earned him the title of "The Insects' Homer", found something especially compelling in what has been variously called the *sphex* or solitary wasp, a family of parasitic wasps.[2] His attraction to these creatures is not unique in the history of modern entomology; George and Elizabeth Peckham, for example, have added their work at Pine Lake in northern Wisconsin to Fabre's descriptor of "transcendental" and lodged these wasps in a cultural space of "solitary" as opposed to other wasps and bees that are "social."[3] In ancient China a very different view of these supposedly solitary creatures dominated social discourse about bees and wasps. Far from associating these wasps with being non-communal, ancient Chinese authors held these wasps to be a pure expression of social stewardship; this may be partly due to a possible misunderstanding of their actual habits, but it is also an articulation of a very different—in fact engendering—view of the term "solitary."[4] This "othering" of nature is a nearly universal mode of knowledge production both in the ancient and modern worlds, at times in the form of metaphor or allegory and at others as an organizing principle for observations, or as is most common a blend of these.

This article will focus on the analogical role that the parasitic wasp (in the Eumenidae family or *guoluo* 蜾蠃 in Chinese) and its host the small corn earworm (variously identified as *Gryphodes pyloalis* in the Crambidae family, *Helicoverpa zea* in the Noctuidae family and *Heliothis armigera* also in the Noctuidae family, *mingling* 螟蛉 in Chinese)[5] played in early Chinese discourse. I am particularly interested in the ways in which this parasite/host relationship was used to describe the physiological relationship between teacher and student, sage and disciple, parent and adopted child—all of which were fundamental to the Han era (206 B.C.E.–220 C.E.) Ruist (Confucian) conception of educational practice and moral behavior. Such worm/wasp relationships, as depicted in early Chinese texts, can be used as a tool to reveal nuanced conceptions of language and identity within Han culture.[6]

> The offspring[7] of a *mingling* (corn earworm) as it was dying (paralyzed)[8] met with a *guoluo* (solitary wasp). The wasp prayed over it saying, 'Be like me! Be like me!' After a long period the worm became like the wasp. [But] how fast indeed did the seventy disciples [of Confucius] become like him![9]

Confucius dragged away his paralyzed disciples to his lair and, standing over them, he cast his spell: "Be like me, be like me." In short order, they emerged from the nest just like him. This jarring image, though slightly exaggerated, is from the work of the highly influential poet and thinker Yang Xiong 揚雄 (53 B.C.E–18 C.E.), who observed the relationship between wasp and worm and linked the conduct of the wasp to that of the Ruist sage, Confucius.[10] Our author observes natural phenomena and applies his findings to the human realm, drawing parallels where possible—thus, it is not only a metaphysical relationship, but an *allegorical* one as well.[11] Does Yang Xiong understand the wasp to be a parasite that injects a paralyzing poison into the worm and carries it away to its nest where it injects eggs into its victim's body, thereby using these living corpses as food for young that slowly devour the worm from inside? Or is his understanding more nuanced? What are we to make of his allegory? To answer these questions, I will outline the categories of *chong* 蟲 (small creature) and *feng* 蜂 (bees), wasp, and worm as they were understood in Han Chinese culture. Then, I will return to the image of the wasp and worm to discuss shifts in their allegorical, literary usage as the natural habits of these creatures were discovered, observed, recorded, and disseminated to a broader audience. I am particularly interested in unpacking the Ruist use of the allegory over time, the interplay between image and observation, "testament" and "witness" and the deployment of these tools in the construction of the image of the Ruist sage during the Han dynasty.[12]

Frame: Chong

> There are 300 bare *chong* that have the humans as their master. Accordingly, people are also a type of *chong*. People eat the same food as *chong*, *chong* also eat the same food as people. They are both *chong* eating each other's food—what is odd about this? If *chong* had knowledge, they would also object to people, saying, "You eat that which heaven produces, we also eat it, you say that we are a pestilence, but you do not describe yourselves [thus even though you do the same thing] as a disaster." For all things that breath *qi* (~the enervating material of the cosmos) when they eat what they like, there is no difference in [the way] their mouths and stomachs [react to the food].[13]

This passage, taken from Wang Chong, a name unrelated to the "small creatures," reveals the conceptual framework that enabled the compelling use of the *guoluo* wasp as a literary analogy for the human sage. In early Chinese parlance, such small creatures as the *guoluo* wasp and *mingling* worm are categorized as *chong* 蟲, a large somewhat fluid category of small creatures including but not limited to insects, bugs, reptiles and amphibians. For Chong, a scholar of the later-Han dynasty and a self-proclaimed follower of Yang Xiong's teachings, these non-human creatures were closely linked to the human realm and human conduct. Chong's passage reflects ideas seen in ritual texts, such as the *Da Dai Liji* 大戴禮記, that were circulating in Han intellectual cirucles. In these texts, the category of *chong* is divided into five groups: winged, hairy, shelled, scaled and finally, bare (that is without the aforementioned coverings), each of which is said to be ruled by a great creature such as a phoenix or—in the case of the little bare creatures—the human sage; for scholars and intellectuals, these creatures played a productive role in ritual and cosmological discourse.[14]

In addition to their use in intelectual discourse, visual representations of these creatures were an integral part of the aesthetics of ritual spaces and as such occupied prominent places in the design of ritual accoutrement such as robes, musical instruments, furnishings and tools.[15] The iconographic status of *chong* was at the same time linked to their physical presence in the natural world and their appearance and disappearance at specific times in the ritual and agricultural calendar of early China. *Chong* were understood to have a very close relationship to the eight winds, which were physical and metaphysical constructs enervating the world and informing such broad subjects as geography and cosmography as well as narrow questions of health and medicine.[16] The life cycles of *chong* not only marked the changing of natural and ritual seasons, but also political seasons; infestations of *chong* marked a ruler's poor moral and ritual standing, and are ranked among disasters such as floods and earthquakes. These small creatures, therefore, were believed to be closely linked to the enervating forces that animate reality and thus became even more potent as allegorical images, thereby occupying a prominent place in philosophical, cosmological, religious, and political language of early China.

Breadth: Bees

> Bees are indicative of the relationship of duty between lords and vassals… If one wants to speak of loyally dying for the sake of the principle, of making a name or teaching, none is greater, and so I can not but make a record of this and compose this *fu* to the bee.[17]

In China *guoluo* wasps were included in the larger family of *feng* 蜂 (bees), a category subdivided by their ability to produce honey, their physical shape, and their nesting habits. In fact these divisions overlap, as the honey producing bees were "large" and often made nests; those that did not were "narrow waisted" and lived in the ground or trees.[18] These are categorical differences that are based on a mix of observed and cosmological knowledge, in which their habits are always seen as analogues to human society.

As materials, bees, their larvae, and the honey they produced were seen as foodstuff as well as medicinal components.[19] Honey appears in written records no later than the late Warring States period (481–256 B.C.E.) and was cultivated since at least the 3rd century C.E. when it is recorded that the recluse Jiang Chi 姜岐 (147–188) specialized in raising bees.[20] Wang Chong in addition to a general discussion of *chong* also discusses bees; he categorizes them with fish and birds since they are born from eggs, but then groups them with *fu* 蝮 (viper), *she* 蛇 (reptile), and *chai* 蠆(scorpion) since they are all among the "creatures that are born by embodying overwhelming amount of *yang qi* [a substance related to action and aggression], and all have poison".[21] It must be noted that this grouping reflects a much older discussion of bees/wasps and scorpions as the epitome of poisonous creatures, dating back to the Warring States period.[22] Furthermore, these two groups are subcategorized based on the intrinsic *yin* or *yang* quality where they are born: fish, vipers, and reptiles are born in grassy low moist *yin qi* places, while birds, bees and scorpions are born in nests/hives (*wu* 屋) and trees, which all are elevated *yang qi* spaces. Simultaneously these categories are crossed, fish and birds are categorically the same (*tong lei* 同類) because of egg laying, while snakes and bees have the same nature (*xing* 性) because of their poison and are also categorically the same.[23]

Wang also notes that in the human realm, a type of person that is overwhelmed by *yang qi* is a *xiao ren* 小人 (petty person). Just as they themselves are *yang wu* 陽物 (*yang* material), the honey of bees is *yang* and must be consumed with restraint since, unlike other sweets which are good for the body, this overwhelmingly *yang* material is not when taken in too great a quantity.[24] Wang goes on to describe how it is the case that this exudation is used to cure illnesses that are *yang* in nature, such as those caused by ghostly possession; since he categorizes ghosts as *yang* in nature his cure for them is also *yang*.[25] This vision of these creatures was highly influential and it was not until the Song era (960–1279) gazetteer *Hengxi san shan zhi* 淳熙三山志 that we find detailed accounts of the life of bees.[26]

Another striking feature that links the *yang* aspect of this creature to the physiognomy of the human realm is the manifestation of patterns (*wen* 文), something that is deeply linked to the physiognomy of the sage.[27] It might be tempting to take these as simple equivalencies, but as is clear from the organization of relationships noted above, these are complex collections of features that are not binary or even dualities but are multi-scaled and simultaneous.[28] What is underlying this link between wasp and sage in the mesh of features is the potency that the *yang qi* imbues, a potency that carries over, though not directly, to the vision of the masculinity that is embodied in the wasp and worm imagery, as we shall discuss below.

The Wasp

> The narrow waisted [wasp] has no female counterpart and it is a type of bee. [Since it has no females] it takes the offspring of the silk worm or grasshopper, which it cherishes turning them into its own offspring.[29]

As for this particular kind of wasp (narrow waisted), there is a range of terms in classical Chinese, such as *pulu* 蒲盧, *zhenchong* 貞蟲, *yeweng* 蠮螉 and so on, each term tracing its definition through the commentarial tradition back to the *guoluo*.[30] In the modern biological sense these are not likely designating the very same creature down to the species level, indeed, it is also possible due to physiological similarities that certain large parasitic flies such as robber flies (*asilidae*) were also counted as narrow waisted wasps.[31] Some of the common features associated with these creatures in early China are: they are only male, reproducing only through adopting the larva of other creatures such as moths, spiders or corn earworms which are then transformed into wasps.[32]

There are numerous pre-Han references to this type of wasp, as broadly defined, but at their core is the concept of *hua* (transformation), the foundation of the analogical and biological understanding of this creature that predates the purely Ruist usage by Yang Xiong:

> After not going out for three months, Confucius went back to see [Laozi] and said, "I get it! The magpie hatches its young, the fish spits out its young and the narrow waisted wasp transforms them, and when the younger brother is born the older cries. For a long time I have not myself taken my place in relation to transformation! If I myself do not take my place in relation to transformation how can I transform others?"[33]

This passage is from the "Tianyun" 天運 chapter of the *Zhuangzi* 莊子, a Daoist canonical work traditionally attributed Zhuang Zhou 莊周 (369–286 B.C.E.). This section dated to ca. 180 B.C.E., well before Yang Xiong, equates this creature with *hua*, the transformative power of the natural world. The wasp has the power of *hua* and so can transform others into copies of itself, resembling the sage who internalizes this cosmic power of transformation to transform others.[34] This type of "reproduction" sets these two apart; the wasp and the sage, in their respective arenas, have the power to leap categories extending internal transformation beyond the bounds of their own bodies. It should be noted that this is not meant as a metaphor, both are examples of the same overpowering masculine force. Though this passage appears in a Daoist source, we still find the figure of Confucius as a connection to Yang Xiong's usage of this association, which remind us of the broad spectrum of what would become orthodox Ruist sources.[35]

This physical potency, that both the Wasp and the sage possess, in part dictated by their physiognomy, means they are among what I term "trans-generative creatures." These creatures reproduce through such processes as transformation, metamorphosis, spontaneous birth, conception/births through encounter (usually with the raw or excessive *yin* or *yang* force).[36] This group includes creatures of one gender, hermaphroditic creatures, those generated by water such as the dragonfly, and certain special cases in the human realm such as the famous birth of Houji 后稷 who was born to Jiang Yuan 姜嫄 after she stepped in the toe print of a giant foot print, and so on. In other words, these are all forms of generation that do not employ sexual reproduction.

The Worm

> The anomalous phenomena of a state in chaos are reflected by the swarming of *ming*. This will be an unfortunate year.[37]

As for the *mingling* worm, as we have noted above, it is the larva of a type of moth that feeds on the core of grains such as rice, wheat, and millet. As a larger category such creatures were called *ming* 螟 or *ming chong* 螟蟲 and were for the most part categorized in early Chinese texts as *hai chong* 害蟲 (pests) along with *te* 螣 (locust), *mao* 蝥 (tiger beetle), and *zei* 賊 (stemborer worm); each of these four occupying a different space on the plant, the *te, mao*, and *zei* eat from the outside respectively attacking the leaves, roots, and joints; the *ming* attacks from the inside eating the core.[38] These creatures are discussed in the very particular discourse of regular seasonality, in the years the worms flourish the state does not flourish for they cause famines that lead to unrest. Such troubled times are noted in the early records as "ming" years.[39]

The Wasp and Worm

Yang Xiong's use of the wasp and worm imagery was just one stop on the long journey of an allegory that murned and churned through the literary record of China from the *Shijing* (*Book of Songs*), one of the earliest literary records dating to between 1000 B.C.E. and 450 B.C.E., to the twentieth century and the striking imagery of the novelist and scholar Lu Xun 魯迅 (1881–1936); along the way it took on different roles while maintaining a certain resonance with earlier prototypical examples.[40]

The wasp and worm allegory is often traced back to the *Shijing* poem

"Xiao wan" (also read as "Xiao yuan") 小宛 which has traditionally been interpreted as political allegory; the line in which the wasp and worm appears is located in the middle of the stanza:

1. In the middle of the plain there are soybeans,
2. Everyone gathers it.
3. The *mingling* worm has a child
4. Which the *guoluo* wasp carries off.
5. Teach and instruct your children,
6. Their fortune will be like yours.[41]

What the wasp actually does to the worm is unclear in this passage: are they adopted and raised by the wasp, locating the parallelism between lines 3–4 and 5–6? Or, as a number of modern scholars have argued, the parallelism is actually between lines 1–2 and 3–4, thus implying one should gather and add to the storehouse just like the wasp by taking away the pesky worm that eats our grain?[42] Given the overwhelmingly negative place that the *ming* worm has in early Chinese discourse it seems fairly convincing that the authors of the *Shijing* poem, given their much closer proximity to agriculture than much of the subsequent tradition, understood that the wasp was parasitic and was using the worm to feed its young and in fact meant by these lines that with the help of the wasp the crops would be plentiful and thus provide for the store house. Either reading is possible in the larger context of this poem, but it is not until the 6th century C.E. that Chinese records describe this wasp as parasitic. It is quite likely that both of these understandings were current from a very early period, though the interpretation that the wasp transforms and teaches the abandoned young took hold in the Han, remaining as the orthodox reading all the way into the 20th century.[43]

In the early Han dynasty, the famous Ruist Dong Zhongshu's 董仲舒 (179–93 B.C.E.) deployed this allegory to back a very specific legal argument around a case of adoption recorded in *Chunqiu jueshi* 春秋决事 (*Judging cases by the Chunqiu*):

> Zhongshu judged: A was without a son, and after he rescued and raised B, even though he was not born to him, for whom would he exchange him? As the *Shi* [*jing*] says: "The mingling worm has a child, the *guoluo* carries it off." The Chunqiu's standard is "the father screens the son." A fittingly hid B.[44]

This reading abstracts the allegory from the agricultural milieu that likely influenced the composition of the poems in the *Shijing*, and relocated it in an urban setting of adjudicating domestic law, forming a longstanding understanding connection still seen in the modern Chinese term *minglingzi* 螟蛉子 meaning adopted child.[45] This use deepens the sense of identity of this allegory, moving from the *Shijing* poem where the poet deploys it as a model for proper action (or alignment of the human and natural realms) to this legal context for determining the exact relationship between father and son, a core relationship in Ruism.

The next expression of this allegory comes in the ritual text the *Liji* 禮記, indeed from one of the most influential sections of that text, the *Zhongyong* 中庸 or *Doctrine of the Mean*:

> Lord Ai [of Lu] once asked [Confucius] about governance. Confucius said, "As for the governance of King Wen and King Wu it is laid out in the records. While they lived, their practices of governance dominated, but when they passed away these practices ceased. The way of the people lies in the thorough application of these practices of governance, just like the management of land lies in the careful managed growth of the vegetation; the management of a state is like a *pulu* wasp. Therefore the governance of the state lies in the people [who manage it], the obtaining of [good] people [as subjects] relies on the body [of the ruler], the refinement of the body relies on the Way, the refinement of the Way relies on humanity, and humanity really means being the person. Taking those closest to you as close is of great importance. What is called principle really just means befitting. Promoting those who are worthy is very important, [following] the degrees of closeness and grades of worthiness; this is the product of *li* (performance).... Therefore the Lordling can not but refine his body, in longing to refine the body one can not but act on the basis of closeness; and when he longs to act on the basis of closeness, he can not but understand people, and in longing to understand people, he can not but understand heaven...."[46]

Here the focus is on the wasp and its power to transform, that is to say it focuses on the parasite over the host, the ruler over his subjects and society in general. When Duke Ai asked Confucius about *zheng* 政 "governance," Confucius marches out the examples of the sage rulers Wen and Wu followed by a description of how the governance of the people is like managing the growth of plants, one must order the lands and plant the rows, while at the same time just like the *pulu* wasp one must select good managers who will embody the vision of the ruler (and perhaps kill the pests that harm the plants). This reading does not dim the nurturing associated with this creature, but in this account the wasp plays double duty: representing the external nurturing of the state and internal cultivating of the ruler. In many ways this case focuses on the internal side, for the "other," the *mingling* worm, is not really being given presence. The *Liji* passage uses the term *pulu* 蒲盧, a term understood very early to mean this sort of wasp and this passage has been nearly universally read to be speaking of the *guoluo* wasp for nearly a thousand years.[47]

We begin to see criticisms of the veracity of this interpretation of the term *pulu* in *Liji* by such Song scholars as Huang Zhen 黃震 (1213–1280), who gives an expansive discussion of the traditional understanding before stating the current understanding. Huang, though opposing a long tradition, develops on his predecessor Zhu Xi 朱熹 (1130–1200), who included this *Liji* passage in his highly influential anthology *Si shu* 四書 or *The Four Books*. Zhu Xi states: "According to Shen Kuo 沈括 (1031–1095) this (*pulu*) is actually *puwei* 蒲葦 (pampas grass), and he is correct in this."[48] Then Zhu Xi goes on to explain that it is because this sort of grass spreads quickly and easily, just like the power of the ruler over the people, that it is used here. Huang's more developed discussion stands out because his rejection of the traditional reading likely involves the first instance of the term *ji* 寄 (to rely on/ live off of/parasitize) to explain the wasp symbolism, discussing the parasitic nature of the wasp in relation to the *mingling* worm as described in the *Shijing*. He actually argues that the original interpretations of the poem based on this understanding and all the subsequent confusion originated with a mistake by Yang Xiong.[49] This is a rather interesting shift away from the allegory of the wasp. It is likely that Zhu Xi and Shen Kuo, as is clearly the case with Huang Zhen, had become aware that the wasp was parasitic, and so felt compelled to subvert its use. Fascinatingly, this new reading of *puwei* still echoes the nuances that Yang Xiong emphasized in his discussion of the *guoluo* wasp, especially the speed of transformation, a detail not present in the original *Liji* or the *Shijing* texts. Zhu Xi's reading of the *Liji* passage subsequently became the standard, and even to this day most modern commentaries and translations read it as *puwei* without note. Setting aside the fairly convincing philological argument for this reading, it erases much of the life of the text.[50]

Zhu Xi, contrary to the above description, maintains the importance of the wasp and worm allegory following the traditional reading of the *Shijing* poem as dealing with adoption and transformation. The later Huang Zhen more thoughoughly rejects all uses of this allegory becuase of he understands the true parasitic nature of the wasp.[51] Clearly for Zhu Xi the weightiness and ideological capital of this vision of teaching is hard to resist, if he did actually know of the parasitic nature of the wasp; furthermore in the context in the *Shijing* there is less room for philological maneuvering since uses the term *guoluo*, as opposed to the less standard *pulu,* matched with the *mingling* worm. His reading in the *Liji* case reveals a way to circument the problem of confronting factual uncertanties of the world described in the Classics while maintaining the valued features of the allegory, and in the case of *Shijing* focusing on the extra-factual aspect of truth based on this form of coordinate relationships.

To return to the interpretation that enervated this *Liji* passage for much of its pre-modern history, scholars read this allegory to reference the power of the sage, or in this specific case the sage ruler, that transforms, materially with their very body their subordinates in a manner correspondent to their role in society. This example, which should be dated to before Yang Xiong, shows how Yang took this older vision of the sage and adjusted the scale from one where the actions of the sage ruler dispersed transformative force across the whole state down to one where the scholar-sage transformed a core of disciples whom he personally "prayed" over.[52] This intensifies the imagery and gives it specific dimensions in practice.

Some centuries after Yang Xiong, the poet Liu Ling 劉伶 (221–300 C.E.) breaks the use of this image wide open: in the closing line of his "Jiu de song" 酒德頌 (In praise of ale virtue) he clinches his argument about the potency of ale to move the author to an altered state that blunts the politico-ritual violence of his era, "Two fellows standing there side by side, like the wasp and worm" demonstrating that the wasp and worm can be used not as just as father and son or sage and disciple but companions holding each other up.[53] In the Tang, Jing Kuo 敬括 (?–771) composed a poem called the *Pulu fu* 蒲盧賦 (*Fu to the pulu wasp*) which begins by drawing on the *Liji* deployment with a focus on *zheng* 政 (governance), but then shifts to a description of how two things with different material bodies can be changed through the intervention of *qi*, and finally comes back and makes clear references to Ruist practices that resonates with Yang Xiong's use. Here the perspective is from that of the wasp, focusing attention on the ruler and sage as parasites.

Some three centuries later, Ouyang Xiu 歐陽修 (1007–1072) composed a poem, the *Mingling fu* 螟蛉賦 (*Fu on the mingling worm*), which rotates the perspective to that of the host: the worm, the position of the child, in a sense giving more agency to the object of the incantation. He allows that the force of transformation is not purely in the parent, but the child has an active if secondary role to play; the volition/capacity, success or failure, is in their body. This deployment is still not disrupted by what must have already been fairly common knowledge that the *guoluo* wasp was parasitic, drawing instead on the force of the two founding deployments of this allegory, the *Shijing* and the *Fayan*. The poem begins with a preface composed by the author:

> The *Shijing* has the line, "A worm has a child and the wasp carries it away." This speaks of things that are of different sorts, something which Yang Xiong addressed in his *Fayan*. Oh, the worm is just a *chong*, it does not have a filial heart, but it is able to take on the bequest of a different sort and become its child. Their wings and hair, shape and nature do not match, they so are different. Now, as for people, a father and mother gives birth to a child, they exhaust themselves in raising and caring for him, never considering him as different from themselves. Thus when a child is unable to carry on the accomplishments of his forefathers, the child of a Ruist will in the end turn to being a merchant, a child of a common family becomes a slave. Oh, how different is the worm! Therefore I have composed this poem to the worm.[54]

Throughout this poem Ouyang emphasized the investment in the oral nature of Ruist teachings drawing on one of the most important contributions that Yang Xiong made to this vision.[55] What is most striking here is that Ouyang has an inverted sense of wonder; Yang Xiong wondered at how much more completely the human sage could transform his disciples compared to the wasp, but for Ouyang the human practitioner often fails, the son often loses his rank and standing, unlike the worm which seems always to replicate exactly his "father." Indeed this is indicative of the fundamental societal changes that had occurred by the Song dynasty. We see a shift from the gathering of sage and disciples to the domestic scene of the father and son, a shift from passing on the Way to passing on social standing. These new usages of the wasp and worm allegory not only indicate these broad societal changes, but also indicate the type of writing this imagery began to inhabit, perhaps having lost some of its philosophical luster as the parasitic nature of the relationship reemerged in common understanding.

Depth: The Sage

> [In comparison to the wasp] he, the master, the ablest of them all, an expert in the niceties of anatomy; he who, with magnifying-glass and scalpel, examined the whole entomological series, leaving no nook or corner unexplored; he, in short, for whom insect organism possessed no secrets can think of nothing better than an antiseptic fluid to give at least the semblance of an explanation of a fact that leaves him confounded.
>
> I crave permission to emphasize this comparison between animal instinct [of the wasp] and the reasoning-power of the sage in order the better to bring to light, in due season, the overwhelming superiority of the former.[56]

Shifting back to Han dynasty, let us return to the influential passage in Yang Xiong's *Fayan* and the beginnings of the Ruist sage's image construction. Here, Confucius draws on his power as a sage, a trans-generative power, to alter his brood of disciples.[57] What makes this revolutionary, as we have noted above, is that after the wasp takes hold of another creature's young, it speaks over it, Yang uses the term *zhou* "to pray or curse," which locates the point of contact between the wasp and the worm, between the sage and his disciples, in oral performance; "be like me! be like me!" he incants. It should also be noted that for Yang, he is not really saying, be "like" me, the phrase *lei wo* is much stronger than "like," perhaps "be me" captures more of the weight of two physiologically different creature becoming identical. For *chong*, the wasp and the worm, identity is in the outer appearance, but for the human realm it is in the inner body, in the heart. The oral nature of this trans-generation is part of a broad understanding of how language works, for Yang Xiong language is a physiological construct that is located in the heart, the organ which both exudes spoken and written language: Yang states "speech is the sound from the heart; writing is the figures from the heart."[58] This is fundamentally a bodily and material function. In this same passage, Yang states that speech is fleeting and limited in the distance it can travel, while at the same time writing can pass across time and space. The leap here is that the wasp-sage gets under the skin of his students and writes himself on the heart of the listener, trans-generatively; yes it is oral, but by their very proximity the potency of the sage imprints, as a teacher, his image directly on their hearts and they become him, with this final result revealed in other sections of the *Fayan*.[59] Indeed, the understanding of the leap actually relies on the wasp-worm imagery for clarity.

Several hundred years after Yang Xiong's deployment, the Three Kingdoms era (220–280) scholar Lu Ji 陸璣 gives his own rendering of this allegory which adds to the basic details discussed variously above, "[The wasp] takes the mulberry worm and carries it off to a hollow spot in a piece of wood or among the wood writing boards and brush pots. Then after seven days it is transformed into its own offspring. According to legend it actually prays over it saying, 'Appear as me! Appear as me!'"[60] This account deepens the imagery by locating where the wasps tends to nest, notably in the very materials that represent the proper place of a Ruist scholar, the very men who are key in the redeployment of this allegory. By citing the same understanding that Yang Xiong advocated, Lu is making a direct connection with the orthodoxy of the Classics, which is the basic topic of his work. Indeed these materials, the writing board and brush pot, are images both used to define Chinese culture from an early period onward, and since at least the Han dynasty to define the Ruist scholars as Yang Xiong's own biography attests; these are the instruments of the sage that allows his teachings to be replicated and embodied through time and space.[61] Indeed, the blending of the oral and the written seen in this imagery must be understood as part of the ritual-political performance of Ruism, as holy items in the performance of morality and social order.[62]

It is clear from our forging discussion that the depth and number of uses of this allegory engendered a certain momentum and level of cultural currency that added to its replication and expansion. As noted variously above, there were some voices of resistance, in part due to a change in the understanding surrounding the facts of the relationship between the worm and the wasp, though I have raised doubts about the veracity of this explanation, or maybe more resulting from a changing vision of Ruism and the role that past scholars were to play in this new formulation of Ruism.

The first influence leading to this shift came after the wasps' nests were actually broken open, and the worms were seen to be infested with "grain like" eggs. The earliest clear expression of this is in the 6th century by Tao Hongjing 陶弘景 (456–536) in his *Mingyi bielu* 名醫別錄 (*Informal Records of Famous Physicians*).[63] Thereafter, we find fairly regular expressions of this, from the very expansive discussions by the scholar Lu Ji discussed above to the Song era and Zhu Yi 朱翌 (*jinshi* ca. 1115) who confirmed the observations of Tao and described how the tradition was just mistaken regarding the wasp and worm.[64] The case of Lu Ji is illustrious of a common approach to this new knowledge, very similar to Zhu Xi's differing readings noted above, while acknowledging the actual matter as it stood on the one hand, he maintained the potency of the imagery for the fundamentally more important task of learning from the Classics, and for the practice of Ruism. This is very much like the various moves made in the modern era to allow the *Bible* to jibe with scientific observation: sometimes twisting and turning either the facts or the content of the *Bible,* or sometimes ignoring the facts for the sake of the greater lesson, testament or witness.

What is important is that the relationship of allegory to fact is not equivalent to the simple binary of faith over reason, but complex multi-scaled deployments that allow for simultaneity. The parasitic wasp most elegantly demonstrates this: "Hosts and parasites are always in the process of passing by, being sent way, touring around, walking alone. They exchange places in a space soon to be defined."[65] Just as the relationship between wasp and worm rotated through time, each located within a set orbit of tradition, their paths crossed and shifted as time and understanding of the tradition changed. Widening the scale, the relationship between humanity and *chong* in early China has a similar symbology and simultaneity in their textual and aesthetic production; as we have seen in such works as the *Lunheng* and *Da Dai Liji*, humanity and *chong* are in nearly every case paired, playing varying roles as parasite and host. On the one hand, they are measured against each other in the deployment of social norms. But on the other hand, at an even larger scale both humanity and *chong* cohabitate and infest the larger cosmological system, both interacting in and showing their reaction to the same collective stimulus, which is fundamental to understanding how this imagery worked discursively.

Exit

To conclude, it is clear that the wasp and worm imagery held a deep power for early Ruists, representing the trans-generative physiological relationship between the sage and his disciples and was not fully reliant on observed knowledge that was alloted a limited space in the construction of truth. The expanse between these two shifted through time creating new gaps for productive employment of this imagery; gaps that were produced through the interplay of resonance and difference between observed knowledge and ideology. Furthermore, we see a shifting of perspectives through time between parasite and host, between testament and witness in a spectrum of connectivity; the modern understanding of the waps' venom, once thought to be a simple bodily fluid, reveals that the relationship is not just between two simple aspects but is influenced by simultaneity, for the venom used to paralyze the worm turns out to be a parasitic virus itself thus further questioning the boundaries of individualized bodies.[66] Though there is no explicit evidence that these wasps were known to be parasitic until 6th century C.E., it is possible that both understandings were available at least before the Han era with the parasitic aspects either being lost in the changing relationship of the elite to agriculture or lost currency and remained as a competing but unconvincing interpretation until further changes occurred.

This allegory is also illustrative of how Ruist educational practice was understood as physiological, located in material space; education was a transferal of the representations from one heart onto another, within the intimate geography of both oral and textual practice that included not only the texts but also the material accessories of texts, the brush and brush stand. Within the context of early China where time and space were non-transcendental, a milfoil stalk when mediated by Ruist practice allowed for divinatory conversancy, just as a text that was the writings of a sage, the material expression of the heart, through a brush, contained trans-generative potential for conversing bodily across time and individual.

Finally, I hope that I have given some of the flavor that these creatures had in the literary, philosophical, and religious discourse of early China and a few glimpses of them in the Western context, which in themselves hint at a complex of resonances and contrasts.

Post Script

"The nest that we were watching was found, at the end of the season, to contain 4661 wasps in various stages of development, and others that we opened contained two to three thousand. This is nothing to the social wasps of China, where a single household is made up of from fifteen to twenty thousand members; but China is a thickly populated country, and perhaps with wasps as with human beings several families live in a single domicile."[67]

Notes

1. Henry Fabre, *Hunting Wasps*, trans. Alexander Mattos (New York: Dodd, Mead, and Co., 1916), 15-16.
2. John Burroughs, *Field and Study* (Boston: Houghton Mifflin Company, 1919), 129 - 139. Fabre's works on wasps include: *Hunting Wasps* (1915), *Mason Wasps* (1919), and *More Hunting Wasps* (1921).
3. The Peckhams for their part were strong supporters of Darwin's theory of evolution and were influential not only in their work as authors but also in Milwaukee in the establishment of educational institutions the promoted biology (George Howard Parker, "Biographical Memoir of William Morton Wheeler, 1865–1937," *National Academy of the Sciences* 19 (1938): 207–208). Their major work regarding these wasps was *Wasps: Social and Solitary* (Boston and New York: Houghton, Mifflin and Company, 1905), which was originally published as *Instincts and Habits of the Solitary Wasps*, in *Wisconsin Geological and Natural History Survey* 2.1 (1898).
4. For the earliest Western observers of these wasps such as Aristotle, these creatures were truly challenging and baffling when it came to fixing a biological gender (*Generation of Animals*, 759a.8–761a.11). It should also be noted that Aristotle (*History of Animals*, 623b.6–13) actually organizes such creatures as bees and wasps with the term "gregarious" and "solitary" of which there are six kinds that are of the former (including bees and wasps) and three kinds of the latter (including the small siren, large siren, and bumble bee). It is indeed this ambiguity or complexity of gender that seems to have increased their circulation, just as we shall see in the Chinese tradition, which might be decisive about their genders (at least as defined in traditional Chinese terms) but affords them very special processes.
5. On these identifications see Hu Miao, *Shijing de kexue jiedu* 詩經的科學解讀 (Shanghai: Shanghai Guji Chubanshe, 2007), 546; Li Haixia 李海霞, *Hanyu dongwu mingming kaoshi* 漢語動物命名考釋 (Chengdu: Ba Shu Shuju, 2005), 595.
6. Within the structure of this paper I take Ruism as a belief system that was in its formative period as an identity during the Han Dynasty (the main focus of this paper), but as with the paper as a whole my use is broad and inclusive since the discussion will span the periods before the Han and more than a thousand years after as we follow the wasp and the worm on their cultural journey. Furthermore, I do not mean to imply that "Ruist" was a fixed form of identity in contrast to Daoist or the like in the Han.
7. It is unclear the exact understanding of the Chinese at this time, but it seems likely that they believed that this was a "child" a small version of a larger creature, not a larva of a moth as we now know it to be, so I have maintained the understanding of the time in my translation.
8. As shall be elaborated below, this is the true state of affairs, the wasp injects its host with a poison that paralyzes it, but it is unclear if Yang Xiong knew this, and if he did, if he was speaking of this.
9. *Fayan* 法言(*Exemplary sayings*), in *Sibu beiyao* 四部備要 (*SBBY*), 1.1b/1.5.
10. Yang Xiong has traditionally been seen as the inheritor of Confucius's teachings after Xunzi. Yang Xiong is known for the bizarre combination of engaging both in strong

Ruist orthopraxis and in the utter radicalism of making anew the venerated works of the past—a position that made him at the same time venerated and vilified, a member both of the exclusive club of inheritors of Confucius's teachings and the even smaller club of loathed usurpers. Mark Pitner, *Embodied Geographies of Han Dynasty China: Yang Xiong and His Reception* (Ph. D. diss., University of Washington, 2010), David R. Knechtges, *The Han Rhapsody: a Study of the fu of Yang Hsiung, 53 B.C. –A.D. 18* (Cambridge [Eng.] ; New York: Cambridge University Press, 1976), 1–11; Wang Qing 王青, *Yang Xiong pingzhuan* 揚雄評傳 (Nanjing: Nanjing Daxue, 2000), 71–73.

11. I have some reservations about this as well but in this context it is functional. Regarding such questions in sinology see: Pauline Yu, "Allegory, allegoresis, and the Classic of Poetry," *HJAS* 43.2 (1983): 377–412; *The Reading of Imagery in the Chinese Poetic Tradition* (Princeton: Princeton University Press, 1987).
12. It should be noted though this study will deploy materials from a broad range of time periods, it must be kept in mind I am not implying a homogeneity, but a complex set of changing relationships and borrowings that are informative of each other to the modern researcher.
13. Wang Chong 王充 (C.E. 27–ca. 100). *Lunheng jiaoshi* 論衡校釋, in *Xinbian Zhuzi jicheng* 新編諸子集成 (*XBZZJC*) (Beijing: Zhonghua Shuju, 1990) 16.49.716. For a full translation of this chapter see Alfred Forke, *Lun-Heng, part II, Miscellaneous Essays of Wang Ch'ung* (Berlin: Georg Reimer, 1911), 363–368.
14. The context of this citation is also very important; it comes from the final chapter of this text on ritual called "Yi benming" 易本命 (Change as the basis of life), which begins by linking the birth of humans with that of the rest of the material world, then proceeds to move across the body and land linking the process of birth, growth and change both physiologically and cosmologically between the larger categories of *ren* 人 (humanity), *qinshou* 禽獸 (birds and beasts), *wanwu* 萬物 (the myriad of things), and the *kunchong* 昆蟲 (swarming little creatures). *Da Dai Liji*, *Sibu congkan* 四部叢刊 (*SBCK*), 13.9A. For a detailed discussion on the cosmological implications of the numerology involved see Zheng Jixiong 鄭吉雄, "Zhongguo gudai xingshangxue zhong shuzi guannian de fazhan" 中國古代形上學中數字觀念的發展, *Taiwan dongya wenming yanjiu xuekan* 臺灣東亞文明研究學刊 2.2 (Dec. 2005): 137–174. Joseph Needham and Wang Ling, *Science and Civilization vol. 2: History of Scientific Thought* (Cambridge: Cambridge University Press, 1956), 232–244, 247–261; Marc Kalinowski, "Technical Traditions in Ancient China and *Shushu* Culture in Chinese Religion," in *Religion and Chinese Society, vol. 1* (Hong Kong: Chinese University Press, 2004), 223–248.
15. *Zhouli*, *Shisanjing zhushu* 十三經注疏 (*SSJZS*) (Taipei: Dahua Shuju, 1982), 41.286.924c–287.924a; K. C. Chang, *Art, Myth, and Ritual: The Path to Political Authority in Ancient China* (Cambridge, Mass.: Harvard University Press, 1983), 56–80; Zhang Anzhi 張安治 and Xue Yongnian 薛永年 eds., *Yuanshi shehui zhi Nanbeichao huihua* 原始社會至南北朝繪畫, in *Zhongguo meishu quan ji* 中國美術全集, vol. 1 (Beijing: Renmin Meishu Chubanshe, 1986), 1–15; Zhongguo meishu quanji bianji weiyuan hui 中國美術全集編輯委員會 eds., *Hua xiang shi hua xiang zhuan* 畫像石畫像磚, in *Zhongguo meishu quan ji*, vol. 18. Shanghai: Shanghai Renmin Meishu

Chubanshe, 1988), 20–21, 109–147; William Watson, *The Arts of China to AD 900* (New Haven: Yale University Press, 1995), 25–26, 89–95.

16. *Shuowen jiezi gulin* 說文解字詁林（*SWJZGL*）(Taipei: Dingwen shuju, 1997), 10, 796, 1002, 1015; *Er ya* 爾雅, in *SSJZS*, 3.714.2640a; *Dadai Liji*, in *SBCK*, 13.7b; "Xici zhuan" 繫辭傳, in *Zhouyi zhengyi* 周易正義, in *SSJZS*, 8.76.88c. Zou Shuwen 鄒樹文, *Zhongguo kunchong xueshi* 中國昆蟲學史 (Beijing: Kexue Chubanshe, 1981), 26–30. John S. Major, "Notes on the Nomenclature of Winds and Directions in the Early Han," *T'oung Pao* 65 (1979): 66–80; Paul Unschuld, "Der Wind Als Ursache des Krankseins," *T'oung Pao* 68 (1–3, 1982): 91–131; Yamada Keiji 山田慶児, "Kyuukan happuu setsu to shou shiha no tachiba" 九官八風説と少師派の立場, *Touhou gakuhou* 52 (1980): 199–24; Paul Unschuld, *Medicine in China: A History of Ideas* (Berkeley: University of California Press, 1985), 67 - 73.
17. Liu Shen 劉詵 (1268–1350) preface to his *fu* style poem titled *Feng fu* 蜂賦 (*Fu on bees*) (in *Quan Yuan wen*, vol. 22 [Nanjing: Jiangsu Guji Chubanshe, 2001] 681.4). This poem seems to be playing on the homophony between *feng* (bee) and *feng* 封 (enfeofment or promotion), which is common by the Yuan dynasty (1234–1367).
18. Yang Xiong, *Fangyan*, 11.11; Gan Bao 干寶(fl. 315), *Soushenji* 搜神記 (Taipei: Li ren Shuju, 1985), 12.146.
19. *Chuci buzhu* 楚辭補注 (Beijing: Zhonghua Shuju, 1983), 9.208; Zhang Hua 張華, *Bowuzhi jiaozheng* 博物志校證, comm. Fan Ning 范寧 (Taipei: Mingwen Shuju, 1986), 110. Kwang-chih Chang and Eugene N. Anderson, *Food in Chinese Culture: Anthropological and Historical Perspectives* (New Haven: Yale University Press, 1977), 30, 57, 67; Bernard Read, *Chinese Materia Medica: Insect Drugs* (Beijing: Peking Natural History Bulletin, 1941), 10–34; Donald John Harper, *Early Chinese Medical Literature: The Mawangdui Medical Manuscripts* (London: Kegan Paul International, 1998), 479.
20. *Gaoshi zhuan* 高士傳 (Lives of High-Minded Gentlemen), in *SBBY*, b.10b. There are some hints to a much earlier dating such as a number of mentions of bees and the possible raising of them in the *Huainan zi* 淮南子 (in *XBZZJC*, 13.963) and *Shanhaijing* 山海經 (Classic of Mountains and Seas) (Guo Fu 郭郛 ed. *Shanhaijing zhuzheng* 注証 [Beijing: Zhongguo Shehui Kexue Chubanshe], 5.6.1, 7.3.11, 5.6.1) but this latter work has a very complicated textual history.
21. Wang Chong, *Lunheng jiaoshi*, 66.23.955–956.
22. *Zuozhuan* 左傳, *SSJZS*, 15.111.1813c; *Guoyu* 國語, in *SBCK*, 15.9.10b. This aspect of these creatures clearly added to their social currency in early cultures, Aristotle also associates bees and wasps with scorpions, snakes and found their ability to sting one of their gender defying qualities. *Parts of Animals* 683a5–25; *On Marvelous Things Heard*, 844b.31–35; *Generation of Animals*, 759b1 - 7; Robert Mayhew, "King-bees and Mother-wasps: a Note on Ideology and Gender in Aristotle's Entomology," *Phronesis*, 44.2 (1999): 127–134.
23. Wang Chong, *Lunheng jiaoshi*, 66.23.957.
24. *Ibid.*, 66.23.951.

25. *Ibid.*, 66.23.951. This sort dance between healing and harm is common in the West as well, though located in the binarity of good and evil, or as recently made famous by Derrida's *Pharmikon* (Jacques Derrida, *Disssemination* [Chicago: University of Chicago Press, 1981], 95–117) located in the differential between the two. Where in China there is not this tension, these aspects are material qualities that just need to be adjusted for, in a sort of simultaneity of state, based on time and place, context and scale.
26. Liang Kejia 梁克家 (1127 - 1187), *Hengxi san shan zhi*, in *Song Yuan zhenxi difangzhi congkan* 宋元珍稀地方志叢刊, vol. 7 (Chengdu: Sichuan Daxue Chubanshe, 2007) 42.1719.
27. Wang Chong, *Lunheng jiaoshi*, 66.23.958.
28. Of course, dualities were a very powerful structure in early Chinese discourse, but were applied in a complex manner especially among such thinkers as Yang Xiong and Wang Chong.
29. Zhang Hua 張華 (232–300), *Bo wu zhi jiao zheng*, 4.45.
30. Additional terms: *ben feng* 奔蜂, *xiyao feng* 細腰蜂, *tufeng* 土蜂, *tucun* 土郸, *zhifeng* 稚(穉) 蜂, *yeweng* 蠮螉, *youtui* 蚴蜕, *gangwa* 缸瓦蜂. Certainly these variations must in part at least reflect regional and species variations, even today plant names and names of specific insects generally have a broad range of variation, but besides Yang Xiong's (*Fang yan*, 10.10) designation of *yeweng* as the regional term for this wasp in the area between Yan 燕 and Zhao 趙 all distinctions are more or less lost to us today.
31. Besides the work of Fabre and the Peckhams noted in the opening section of this article, regarding these wasps. See Roberto Boesi; Carlo Polidori; José Tormos; Stefania Bevacqua; Josep Daniel Asís; Francesco Andrietti, "Trap-Nesting Ancistrocerus sikhimensis (Hymenoptera: Eumenidae) in Nepal: Nest Structure and Associates (Hymenoptera: Chrysididae; Acarina: Saproglyphidae)", *The Florida Entomologist* 88.2 (Jun., 2005): 135–140; Jeffrey A. Harvey, Ian F. Harvey, David J. Thompson, "Flexible Larval Growth Allows Use of a Range of Host Sizes by a Parasitoid Wasp," *Ecology* 75.5 (1994): 1420–1428. Margaret Warner Morley, *Wasps and Their Ways* (New York: Dodd, Mead and Co, 1901) esp. part 2.
32. Xu Shen, *SWJZGL*, 10.875–879; Gan Bao (fl. 315), *Soushenji*, 13.164–5.332; trans. DeWoskin, Kenneth J. and J. I. Crump, *In Search of the Supernatural: The Written Record* (Stanford, Calif: Stanford University Press, 1996), 157–8; *Liezi* 列子, in *XBZZJC*, 1.16; trans: A. C. Graham, *The Book of Lieh-ztu: A Classic of the Tao* (New York: Columbia University Press, 1990), 21.
33. *Zhuangzi*, in *SBBY*, 5.27a.
34. Dating based on A.C. Graham, "How much of *Chuang tzu* did Chuang Tzu write?" in *Studies in Chinese Philosophy and philosophical Literature* (New York: New York University Press, 1990), as cited in H. D. Roth, "*Chuang tzu* 莊子", in *Early Chinese Texts: A Bibliographic Guide* (Berkeley: University of California Press, 1993), 56–57.
35. Yang Xiong had studied the *Zhuangzi* at an early stage with Zhuang Zun 莊遵 (Yan Junping 嚴君平) (ca. 83 B.C.E. –bf. 6 C.E.), the Daoist recluse and expert on the *Laozi*, *Zhaungzi* and *Yijing*. Aat Vervoorn, "Zhuang Zun: A Daoist Philosopher of the Late First Century B.C.," *Monumenta Serica* 38 (1988–89): 69–72; Tang Yijie 湯一介, *Yan*

Junping pingzhuan 嚴君平評傳 (Nanning: Guangxi Xinhua Shudian, 1997) 9–11. In addition to what would become understood as Daoist sources, this creature also appears in the *Mozi* 墨子, an early collection of writings that dates in part to a time nearly contemporaneous to Confucius that exhibit a striking skepticism towards Ruism as it stood at the time and the practical use that professional Ruists played in society. *Mozi wengu* 墨子閒詁, in *XBZZJC*, 8.31.238; 8.32.257–8; trans. Burton Watson, *Basic Writings of Mo Tzu* (New York: Columbia University Press, 1964), 103, 113–114.

36. *Liezi*, 1.15–17; Trans: Graham, 21; *Sou shen ji*, 12.146.300; 13.164–5.333; trans. Kenneth DeWoskin and J. I. Crump, 142–4, 157–8; Huang Diming 黄涤明 and Ding Wangdao 丁往道 trans., *Anecdotes about Spirits and Immortals*, vol. 2 (Beijing: Foreign Language Press, 2004), 689–90, 761.
37. Att. Lü Buwei 呂不韋 (d. 235 B.C.E.), *Lushi chunqiu* 呂氏春秋, cited in the "disaster" section of the Ouyang Xun 歐陽詢 (557 - 641) comp., *Yiwen leiju* 藝文類聚 (Shanghai: Shanghai Guji Chubanshe, 1965), 100.1732.
38. The line from the *Shijing* poem *Great Fields* 大田 (Mao 212) reads: "Get out [of my fields] you *ming* and *te*, and you too [foul] *mao* and *zei*" 去其螟螣，及其蟊賊; the Mao commentary goes on to specify where each creature feeds as described (*SSJZS*, 14. 208.476c). Huang Hsing-Tsung, "Plants and insects in man's service," in *Science and Civilization in China*, vol. 6.1, 521–522.
39. "Yin gong wu nian" 隱公五年, *Chunqiu* 春秋 (*Spring and Autumn Annals*), *SSJZS*, 3.24.1726b.
40. Lu Xun 鲁迅, "Po'e shenglun" 破惡聲論, in *Ji wai ji shiyi bubian* 集外集拾遺補編, in *Lu Xun quanji* 鲁迅全集, vol. 8 (Beijing: Renmin Wenxue Chubanshe, 2005) 27.
41. *SSJZS*, 13.183.451 (Mao 196); trans.: Bernhard Karlgren, *The Book of Odes* (Stockholm: The Museum of Far eastern Antiquities, 1950)143–45; Arthur Waley, *The Book of Songs* (New York: Grove Press, 1996), 176–77.
42. Huang Hsing-Tsung, "Plants and insects in man's service," in *Science and Civilization in China*, 6.1, 520–521.
43. This reading of the *Shijing* was cemented by the great late Han hermeneut Zheng Xuan 鄭玄 (127 - 200) who reads this as a metaphor for adoption (*SSJZS*, 13.183.451), then a century later it is reinforced by Lu Ji 陸璣 in his *Mao Shi cao mu niao shou chong yu shu* 毛詩草木鳥獸蟲魚疏 (in *SKQS*, 70.12a). Huang Hsing-Tsung, "Plants and insects in man's service," in *Science and Civilization in China*, vol. 6.1, 523 - 4.
44. Dong Zhongshu, *Chunqiu jueshi* 春秋決事 as cited in the *Tong dian* 通典, 69.29.29.14; Ma Guohan 馬國翰, *Yuhanshan fangji yishu* 玉函山房輯佚書 (Shanghai: Shanghai Guji Chubanshe, 2002) 1202.256a. For a full translation and study of what remains of this work see Michael Loewe, "Dong Zhongshu as a Consultant," *Asia Major* 22.1 (2009): 163 - 182.
45. On this as a metaphor for adoption see Ann Waltner, *Getting an Heir: Adoption and the Construction of Kinship in Late Imperial China* (Honolulu: University of Hawaii Press, 1990), 24 - 27.
46. *Liji*, in *SSJZS*, 52.401.1629b.

47. A list is found in the *Liji jishuo* 禮記集說, in *SKQS*, 120.130.1a–46b; representative are Zheng Xuan who cites the *Shijing* and the *Er ya* and gives the gloss from Yang Xiong's *Fang yan* (*Liji zhengyi*, 60,2012) and then Lu Deming 盧德明 (556–627) in his *Jingdian shiwen* 經典釋文 [(Taipei: Xuehai Chubanshe, 1988) 14.4.3a.209a] both of which gloss this as the *guoluo* wasp.
48. Zhu Xi, *Sishu jizhu* 四書集注 (Taipei: Yiwen Yinshuguan, 1980) 15a. Shen Kuo, *Mengxi bitan* 夢溪筆談, in *SKQS*, 862.3.11b.
49. *Huangshi richao* 黄氏日抄, in *SKQS*, 707.4.49b–50a; 25.37.
50. Legge, *The Chinese Classics* (Hong Kong: Hong Kong University Press, 1960), 1–2.405, Wang Wenjin 王文錦, *Liji yijie* 禮記譯解 (Beijing: Zhonghua Shuju, 2001) 784 - 785.There are a few exceptions like, Yang Tianyu 楊天宇 who reads this as wasp (*Liji yizhu* 禮記譯注 [Shanghai: Shanghai guji chuban she, 1997], 910 - 911).
51 Zhu Xi, *Shijing zhuan* 詩經傳 (Shanghai: Shanghai guji chuban she, 1987), 94.
52. As is clear from the nuances that Zhu Xi exhibits in his deployment of *puwei* in place of *pulu* and the following uses that clearly draw on Yang Xiong's use, though clearly not the originator of this reading, can not fully be released from his place as one of the instigators of this "trouble" as Huang Hsing-Tsung so strangely argued ("Plants and insects in man's service," in *Science and Civilization in China*, vol. 6.1, 522).
53. "Jiu de song" 酒德頌, in *Wenxuan* 文選, 47.8a–9a.
54. Ouyang Xiu, *Mingling fu*, in *Ouyang Xiu quanji* 歐陽修全集 (Beijing: Zhonghua shuju, 2001), 58.838.
55. For a contemporary perspective on Ouyang's use on the *mingling* worm in Ruist teaching see Huang Zhen, *Huang shi ri chao* 黄氏日抄, in *SKQS*, 708.61.21.
56. Here J. H. Fabre is making a direct comparison between the paralyzing capacity of the wasp and the "sage" scientist. Henri Fabre, *The Hunting Wasps*, 16.
57. This is likely the reason Yang has been marked as the main perpetrator of the befuddling of scientific truth.
58. *Fayan*, 5.3b/5.13.
59. *Fayan*, 3.4b/3.21; 2.4b/2.21; 8.4b/ 8.18; 5.1b/ 5.3 and so on.
60. Lu Ji, *Mao Shi cao mu niao shou chong yu shu*, in *SKQS*, 70.12a; Huang Hsing-Tsung, "Plants and insects in man's service," in *Science and Civilization in China*, vol. 6.1, 523.
61. David R. Knechtges, "The Liu Hsin/Yang Hsiung Correspondence on the *Fang Yan*." *Monumenta Serica* 33 (1977-78): 316.
62. Then in the following centuries and millennia these items would be fetishized and maintain their status as markers of the literary elite even as the role and make-up of this elite changed.
63. Tao Hongjing, *Mingyi bielu* 名醫別錄 (*Informal Records of Famous Physicians*), now recorded in Li Shizhen's 李時珍, *Bencao gangmu* 本草綱目 (SKQS, 39.16b); *Needham* 6.1, 524.
64. *Lushi Shi shu guangyao* 陸氏詩疏廣要, in *SKQS xia zhi xia* 下之下 70b - 75a; 猗覺寮雜記, (also found in Beijing: Zhonghua Shuju, 1985) *shang* 上 41; Zhu Yi'an 朱易安 ed., *Quan Song biji. Di 3 bian* 全宋筆記.第 3 編, vol. 10 (Zhengzhou Shi: Daxiang Chubanshe, 2008) *shang* 上 53 - 4.

65. Michel Serres, *The Parasite*, trans. Lawrence Schehr (Minneapolis: University of Minnesota Press, 1982), 16.
66. Nancy E. Beckage in her article "The Parasitic Wasp's Secret Weapon" (*Scientific American* 277.5 [1997]: 82–87) has shown that there are in fact three participants in this relationship, the parasitoid wasp, the worm and a virus; the latter is present in the wasp, if not part of the wasp, "virus and wasp appear to be permanent and integral partners" (86) sharing DNA, and is injected by the wasp into the worm to disable it, in fact the virus is so close to the wasp that the wasp's DNA encodes the genes for both.
67. George W. Peckham and Elizabeth G. Peckham, *Wasps: Social and Solitary* (Boston and New York: Houghton, Mifflin, and Company, 1905), 12.

A SNAKE IN A BASIN AND A WORM IN THE FLESH: EXTERNAL SERPENTS, INTERNAL WORMS, AND AUTHORITY OVER THE BODY IN THE *LEGENDA AUREA* OF JACOB DE VORAGINE

Brenda S. Gardenour

I'd like to begin with a story, which will also be the end of our story, taken from a miracle collection written for the Dominican Saint Peter of Verona, who was martyred in 1252 by axe-wielding heretics while walking along a road outside of Milan.[1] Just before he died, he rose to his knees, recited part of the Creed, and offered his blood as a sacrifice to God; he then dipped his finger in his own blood and wrote "*credo in Deum*" in the dirt.[2] Peter's broken body was brought to Milan by is Dominican brothers, where both his corpse and his cape were enshrined. Soon after his entombment, both relics began to work miracles, many of which were thaumaturgic in nature. So numerous were the miracles wrought by Peter in both life and death, that Pope Innocent IV canonized him just one year after his martyrdom.[3] The miracle tale recounted below was presented as evidence of the saint's divine power to heal:

> Once, there was once a young man named Henry whose throat was almost completely constricted by a tumor, and he could neither speak nor breathe; because of this, his parents worried that he was near death, and despaired of his life. Hearing Brother Peter of Verona was near, they walked quickly and caught up with him on the road, placed the dying child before him, and prayed for his help. Brother Peter raised his hands to the sky and made the sign of the cross, held out his cape, and placed it over the child, after which the child recovered. Soon afterward, the child's father suffered from vehement pains in his stomach and intestines. Fearing his own death, he reverently took the cape once again and placed it across his chest, after which he vomited worms having two heads and covered with dense and vile hair. After this purgation, the man recovered to full health.[4]

In the above miracle, a father's faith, the prayer of Saint Peter, and contact with his holy cape were powerful enough to rupture young Henry's rotting tumor and to purge his father's belly of worms. On the surface, a graphic discussion of bodily effluvia such as pus, vomit, and potentially parasitic worms seems out of place in a text written to glorify a holy man and justify his canonization. Perhaps even stranger, from the modern perspective, this miracle was not only included in Peter's canonization dossier, but also incorporated into the thirteenth-century *Legenda Aurea*, a devotional text meant to inspire piety and wonder to all those who read or heard the miracles therein.[5] The *Legenda's* author, Jacob de Voragine, culled his miracle tales from biblical and patristic sources, *vitae*, miracle collections and sermons; these miracles, drawn from across several hagiographical traditions and historical periods, were meant to bear witness to God's continued engagement with the sublunary world. The unique nature of the *Legenda Aurea*, in which Peter of Verona's thirteenth-century worm miracle is told alongside the miracles of much earlier saints, such as Martin, Sylvester and Simeon, provides us with a snapshot, taken in 1275, of hagiographical sources; through these, we might trace some of the traditions regarding worms, and by extension snakes, in medieval hagiography.

An examination of the *Legenda* reveals several multivalent and interconnected relationships between worm, snake, human, and text. In those tales culled from early hagiographical and patristic sources, snakes and worms serve very different didactic purposes. Snakes, which infest Christian regions rather than Christian bodies, spread both plague and heresy and must be expelled; worms, however, are sent by God to invade Christian bodies to serve as living witness to either the sinful or saintly nature of the afflicted, and because of this cannot and should not be purged. Contemporary with these sources is the story of the swallowed serpent, in which an individual imbibes a snake which then must be flushed from the body; what begins as an imaginary snake in medical hagiography becomes a very real serpent in miracle tales such as those attributed to the heavenly physicians, Cosmas and Damian. In thirteenth-century miracles, the concern is no longer with the expulsion of swallowed serpents, imaginary or otherwise, but instead with very real parasitic worm infestation. While worms in these later texts are still associated with ideas of redemption and punishment, their primary function is to be expelled from the afflicted, thereby curing an otherwise incurable medical condition and proving in very visceral ways the thaumaturgic power of a potential saint. Composed to meet the demands and expectations of late-medieval canonization committees for whom it was written, the worm tale of Peter of Verona is both in accordance with as well as a departure from the earlier Christian and medical hagiographical traditions from which it descended.

Expulsion of Snakes and the Reverence of Worms in Early Miracle Tales

In the hagiographical tradition of Christian late-antiquity, saints were accorded the power of casting snakes and serpents, but rarely worms, from the Christian body, a concept that extended beyond the physical body of a single individual to encompass an entire region and, in some cases, all of Christendom. The *Legenda Aurea* contains several early miracle tales in which snakes and serpents are depicted as toxic invaders of Christian regions. In the miracles of Saint Sylvester, for example, a serpent (*dracone*) dwells in a pit near Rome and slays three hundred men a day with its venomous, plague-inducing breath; descending into the pit, Sylvester stitches the serpent's mouth with thread and binds it with a wax seal bearing a cross, thereby saving Rome from the "venom of the dragon."[6] As the dragon poisoned Christian Rome with his toxic breath, our author argues, so too did pagans and heretics corrupt the city with their lies and false belief.[7] The tale of the venomous serpent whose presence poisoned the good Christians of a region served as a *topos*, a story told again and again. In his sixth-century epistle to Bishop John, Gregory the Great complained that the plague then ravishing Rome was caused by the exhalations of serpents infesting the Tiber River.[8] Saint Hillary, Saint Martin of Tours, and more famously, Saint Patrick were renowned for their ability to purge serpents, both physical and allegorical, from the wild regions where they preached the Christian faith and converted those who witnessed their miraculous control over nature. In each of these cases, it was the divine presence of the holy man and his direct commandment, in the vein of Moses and Christ, which convinced the slithery invaders to leave the Christian body and never return.

Worms, like snakes, are legless invaders and potential sources of disease in the early hagiographical tradition. Unlike snakes, however, worms are not depicted as infesting geographical regions or communities; the worm instead penetrates the human body to become a monster of the microcosm. While several late-antique and early medieval hagiographical sources recount episodes of snake infestation and the miraculous expulsion of venomous serpents, relatively few early sources deal with the expulsion of worms from either regions or bodies. In those sources that do address worm infestation, less emphasis is placed on the nature of the worms themselves than on their divine role—punitive or redemptive—in the body of the afflicted. An example of divine punishment by worms is recounted in the long-told story of King Herod's death, which found its sources in the first-century CE *Jewish Histories* of Flavius Josephus, and came to be incorporated into the medieval hagiographical tradition in various forms.[9] In one version, retold in the section on the History of the Innocents in the *Legenda Aurea*, an aged Herod is afflicted with a fever; as his body begins to putrefy, worms issue from the soles of his feet or, according to other manuscript copies, his testicles.[10] Upon examining his body, Herod's physicians announce that his condition is incurable and undoubtedly the result of God's wrath. For medieval Christians, Herod's death served a didactic purpose; his slaughter of the first born was an attempt to kill the Christ child, a foreshadowing of Jesus' death at the hands of Caiaphas and the Jews.[11] Herod's worms were a sign not only of the corruption of his flesh, but also of his soul, which would be punished accordingly in the world to come.[12]

In early hagiographical sources, worm infestation is a sign not only of God's wrath and punishment, but also of his favor and redemption. One example would be the *vita* of Saint Simeon the Stylite, which was first written in the fifth century CE and ultimately incorporated into later-medieval *flores sanctorum* such as the *Legenda Aurea*.[13] Anthony, the author of Simeon's *vita*, writes that "from the time that he was in his mother's belly he was chosen of God," a status confirmed both in his piety and his austerity. As a young monk, Simeon bound himself with a cord so tightly about the torso that "his flesh rotted," a great stench rose from his body, and his bed became filled with worms.[14] Upon discovering this asceticism, his abbot exiled him from the monastery, and Simeon moved to a nearby pillar of stones, which over the decades was augmented to a height of thirty cubits. After fourteen years upon his pillar, Simeon suffered an aposteme on his thigh; the worms that issued forth from his suppurated flesh fell to the ground and immediately proved to have miraculous qualities. One of these worms, when held by the heathen king Basilike, turned into a precious stone; likewise, worms that fell from the pillar onto the eyes of a blind man immediately restored his sight. Saint Simeon's miraculous worms served as a sign of his sanctity and his status as one of God's chosen disciples. Through his rejection of the body and his willingness to suffer both spiritually and physically, Simeon's flesh and its wormy corrupters became a nexus between the earthly and heavenly realms, a point of contact with the divine.

In both Herod's divine punishment and Simeon's miraculous reward, worm infestation is deeply personal, a reflection of God's judgment rendered in the corruption and consumption of the flesh. Snakes, on the other hand, generally represent an external threat, usually linked with heresy, to the larger Christian body; unlike the divine worm, the heretical and toxic serpent must be cast out for the Christian body to survive. Saint Simeon demanded that the worms which fell to the ground be collected and reapplied to his wound, so that they might continue to devour his flesh; at the same time, he cast serpents from both his community and the bodies of those dwelling within it. In the first instance, a local dragon so venomous and with such noxious breath that nothing grew around his cave emerged one day and made his way to Saint Simeon. There, he placed his eye, which was sore, against the holy man's pillar; after three days, during which the dragon harmed no one, Simeon advised his followers to place soil and water upon the dragon's eye, from which a wooden stake over a cubit long sprang forth. After this miracle, the dragon rose, "adored at the monastery almost two hours," and returned to his cave without causing harm to anyone ever again.[15] In the second instance, a woman who had swallowed a snake, and for whom "diverse medicines and physicians" had proven useless, was sent to Simeon who recommended the application of soil and water to her mouth. At the application of the poultice, a serpent slithered out of her gullet and was immediately slain; the snake's carcass was then hung from the base of Simeon's pillar as a sign that the region was clean, protected by Simeon's sanctity, to which his own personal worm infestation bore witness. Where the divine worm works upon the individual and serves as an unmovable sign of God's judgment, the community serpent, be he dragon or snake, must be slain or otherwise vanquished.

Commanding the Snake: From the Physician's Snake in the Basin to the Saint's Snake in the Belly

The story of the woman who swallowed a snake and was purged of it by Saint Simeon is a conflation of two metaphorical creatures, the external snake and the internal worm. The swallowed snake serves as a *topos* in both medical hagiography, written to prove the wisdom of the savvy physician, and Christian hagiography, written to prove the power of a saint over physical and spiritual ills. While both traditions claim the story as their own, each uses the tale of the swallowed snake for very different didactic purposes. In the medical tradition, the snake in the belly tale can be traced back to Galen's commentary on Hippocrates' *Epidemics*. In this text, Galen relates the story of one of his patients, a woman who suffered from acute pains in her stomach and insisted that her condition was caused by her swallowing of a serpent.[16] After a thorough examination, Galen assured the woman that she did not have a snake in her belly, but she would not believe him and panicked; realizing that her emotional turmoil was exacerbating her stomach pains, Galen determined to use the woman's delusions to heal her. To this end, he obtained a snake from the marketplace and kept it in a bag by his side. He then arranged to give the woman a purgative. Warning her that she shouldn't see the snake emerge from her mouth, he blindfolded her, and as she vomited, he slipped the snake into the basin. When he removed the blindfold from his patient, she saw the snake slithering about in the vomit, and was wholly cured because she believed herself to be so.

The snake in the basin tale is retold in Greco-Arabic medical texts across the centuries. First found in Galen in the second century, it is again attributed to the Byzantine Alexander von Tralles in the sixth.[17] In this retelling, Alexander asks the woman what the snake looks like, and he is careful to engage a hunter to procure one of the correct species. In the ninth century Abbasid Caliphate, al-Tabari includes Galen's story in a chapter on the anecdotes of famous physicians in his treatise, the *Paradise of Wisdom*. The only significant change in this rendering of the tale is the gender of the patient, who has now become male.[18] Rhazes, in the tenth century, also discusses the case of the woman introduced by Galen; while he routinely questioned Galen's authority and argued against charlatanism, he did agree with the therapeutic value of Galen's surreptitious introduction of the snake into the basin.[19] In later sources, the snake in the basin tale would be attributed to Rhazes himself and again to Avicenna, both physicians worthy of hagiographical treatment. While the validity and provenance of the story itself is difficult to discern, the function of the tale in medical hagiography—the glorification of the wise physician—is clear. The snake in the basin depicts a physician who understands the relationship between the mind/soul and the body, and has the wisdom to use one in service of the other. Both Galen in his *On the Errors and Passions of the Soul* and Rhazes in his *Spiritual Physick* discuss the importance of the emotions, linked to mind and soul, in the maintenance of physical health.[20] An emotional imbalance might have a corresponding effect on the humors; inordinate passion, for example, might cause a disease known as lovesickness, with symptoms including an increased pulse, difficulty breathing, sleeping and eating, and despair bordering on melancholia.[21] In the tale of the swallowed snake, the wise physician understands the power of delusion to both cause harm and to cure, and cleverly uses this knowledge to the benefit of his patient.

But what of our snake? From the perspective of the physician, snakes posed a serious danger to human health because of their sharp teeth and toxic venom. Ancient and medieval medical texts, including Hippocratic and Galenic works, the pharmacology of Dioscorides, the *al-Hawi* of Ibn Jazzar translated into Latin as the *Viaticum* by Constantine the African in the eleventh century, the *Almansoris* of Rhazes and the *Canon* of Avicenna, as well as their Latin descendants, contained a chapter on the bites of poisonous creatures such as rabid dogs, scorpions, and serpents. In the *Viaticum*, for example, the section entitled *De Morsu Serpentem* treats the symptoms associated with snake bites, the extraction of snake venom from wounds, as well as the compounding and administration of *theriac*; no mention is made, however, of snakes entering through the mouth and becoming lodged in the stomach.[22] For the physician, it was not the snake that needed to be purged, but instead the worm that wriggled and writhed through the human body. The very real threat of worm invasion was an ancient one, winding its way through sources such as the *Natural History* of Pliny, the Galenic Corpus, and the *Twelve Books* of Alexander Tralles. In medieval medical treatises, the nature, symptoms, and treatment for parasitic worms falls under the heading of *De Vermibus*; in some treatises worms are dealt with under individual body areas, such as *De Vermibus Oculorum*, *De Vermibus Aurium*, *De Vermibus Nares,* and *De Vermibus Ventrem*. Almost always, the treatment for worms was purgation, either through clysters or more commonly, purgatives. The physician, then, concerned himself with the internal worm, not the external or impossibly internal snake.

The swallowing of a snake may not have been a physiological or medical concern for Galen; it was however of great personal concern to the woman whose stomach seemed to roil and wriggle, and who genuinely believed that a legless living creature had invaded her body to poison and consume her. For Galen, the only real snake was the external one swimming in the basin; for the woman, both the internal snake and the external snake were very real indeed. The idea that snakes occasionally slithered into the unsuspecting mouths of those drinking contaminated water or sleeping in open fields and the corresponding necessity to purge the body of that creature became a *topos* in Christian hagiography. It is here that we find the story of the snake in the basin lifted from the shoulders of Galen and placed squarely on those of two other men, Cosmas and Damian, who were not only physicians, but also saints. In this late-antique version of the tale, retold in the *Legenda Aurea*, a peasant has gone out to harvest in the fields and, after a hearty lunch, takes a nap upon the ground; while he is sound asleep, a snake slithers into his open mouth and down to his stomach.[23] Upon awakening, the peasant finishes his task, and returns home to a hearty dinner, after which he finds himself tortured by stomach pains so intense that he cries in a loud voice, "God and Saints Cosmas and Damian help me!" In continuous burning pain, the man runs to the shrine of Saints Cosmas and Damian, and there he falls fast asleep. While he slumbers peacefully, the saints purge the man of a wicked (*malignus*) snake that crawls out through his throat, after which he awakens, completely healed. Looking down upon the exhausted man, a priest announces, "Not by your own hand have you been liberated from the serpent, through which you have come to know what it is to enter into the fires of hell."[24] And saying this, the serpent was no longer visible.

This rendering of the snake in the basin tale differs from that presented in Galen in both the nature of the snake and the didactic purpose of the story itself. In our earlier example, the patient believes that she has ingested a snake, but the physician realizes the impossibility of such a scenario. In this second telling, the patient has no idea what is causing his pain; instead, the divine physicians Cosmas and Damian recognize that the source of his agony is an ingested serpent. This second snake, physically real and present in the man's body, also serves a spiritual purpose. As is in other late-antique miracles, such as those of Saint Sylvester, the snake continues to be a symbol for sin, corruption, and the poison of heresy, thus the priest's injunction that the man's suffering with the snake was but a foretaste of what awaits those who succumb to sin and refuse to believe in the power of God and His saints. Where belief in an imaginary snake afflicted the body of Galen's patient, disbelief or ignorance of the serpent's presence tortured the earthly body of our peasant and might also have endangered his eternal soul. The good Christian must tend to the hidden snakes within and, through God's mercy, purge them and be made whole. Despite these differences, in both the medical and Christian hagiographical traditions the ultimate function of the story is the glorification of the practitioner-healer—whether he be the wise physician who has authority over the human body and mind, or the saintly thaumturge whose divine command over nature grants him the power to cast very real serpents from the mouths of the faithful, much as he purges them from the earth itself. In both contexts, the authority of the physician and that of the saint is made manifest in the successful purgation and subsequent healing of their patient.

The Purgation of Worms: Practical Medicine and Proof of the Divine

In the folios of the *Legenda Aurea*, late-antique and early-medieval miracles rest alongside those written by Voragine's thirteenth-century contemporaries. In these later miracle texts, snakes continue to be powerful symbols for sin, corruption, and the punishment of hell. The thirteenth-century *vita* of Saint Francis of Assisi, for example, has the saint lecturing to his disciples, admonishing them to avoid the evil of money, which is nothing less toxic than a *coluber venenosus*, a venomous snake.[25] While thirteenth-century hagiographers retained the snake as a metaphor, they wrote less of the physical problems long associated with snakes—their infestation of a geographic space, their poisonous bite, and their atypical invasion of the human body—and increasingly turned their attention to the medical problem of parasitic worms. In earlier miracles, worm infestation is a sign of divine punishment or blessing, and cure is either impossible or undesirable. In later miracles culled from the *Legenda Aurea*, however, the causes of worm infestation might be either divine or natural, and the goal is almost always the purgation of worms from the body. In a miracle recounted in the section on Saint James the Less, a worm infestation in the nose serves as a sign of spiritual corruption and provides an opportunity for conversion; in a miracle written of Saint Mark, a natural infestation of worms in a suppurating cancerous tumor provides an opportunity both to reward the devotion of the supplicant and to prove the efficacy of the saint. Our third wormy miracle tells of an impious woman who is afflicted with worms in her eyes as punishment for working on Saint Dominic's feast day. This later tale, written specifically for Saint Dominic's canonization process, combines older themes of divine punishment by worms and saintly authority with the more recent concerns of medical proof and theological authority. The complexities evident in Saint Dominic's worm miracle provide us with some possible reasons for the increased emphasis on worm infestation over that of snakes in later-medieval hagiography.

The worm miracle embedded in the text of Saint James the Less' *vita* begins with the capture of Pilate's servant, Alban, by the governor Vespasian. Upon being questioned as to who he was and where he came from, Alban answered that he was a Jerusalemite, to which Vespasian emitted a cry of joy, noting that Jerusalem was famous for its doctors and surgeons. Vespasian then asked if Alban had any knowledge of surgery, because ever since childhood he had been plagued by a colony of worms living in his nose, which no doctor could cure.[26] Alban said that he knew of a very rare and special medicine, one that had been proven to work miracles. After being warned by Vespasian that his life depended on the efficacy of the cure, Alban told the governor about Jesus and His works, to which Vespasian responded, "I believe that He raised the dead and healed the sick, and that He can heal me." Upon uttering these words, all of the worms fell from his nose and he was completely cured.

This miracle narrative, woven from threads picked up in Josephus, draws on the early hagiographical tradition in which worms signify the internal corruption that results from paganism. Unlike our earlier account of Herod's worms, however, Vespasian is not singled out for divine punishment because of his evil nature or his actions against Christ and His people. On the contrary, Vespasian's worms offer him the reward of miraculous cure and Christian conversion, by which he is healed both physically and spiritually. Vespasian's kinder treatment in the *Legenda Aurea* can be attributed to the medieval view that it was God's will for Vespasian and Titus to siege Jerusalem and destroy the Temple as retaliation against the Jews for the slaughter of Saint James the Less (thus the story's inclusion in this section), the persecution of Christians, and the murder of Christ. Because of his divine role in future events, Vespasian is rewarded with physical proof of God's blessing—the complete purgation of worms. For the medieval reader familiar with medical texts, the eradication of nasal worms would have been miraculous, indeed; according to medical compendia such as Ibn-Messue's *Perfect Medicine*, conditions related to *De Vermibus* were difficult, if not impossible, to cure.[27]

Our second miracle, attributed to Saint Mark, centers on a young Venetian man lying in the street who suffers from a cancerous aposteme in his breast, or in some manuscripts his jaw (*mammilla, maxilla*), which is infested with worms that feast upon the festering flesh.[28] Unable to suffer the pain and horror of his condition any longer, the man prays solemnly to Saint Mark, upon which he instantly falls into a deep sleep. During this incubation, Saint Mark appears to him dressed as a pilgrim, reports that he must go and save the men on a sinking ship, and lays his hand upon the man's worm-infested chest. The man then awakes from his dream to hear sailors exclaiming that their ship would have been wrecked but for the help of Saint Mark. As with Vespasian, it is the man's faith and the power of the saint that expels the worms and heals his body; likewise, there is no evidence that the young Venetian is in a state of extreme sin. In this miracle, worms neither signify punishment nor find their source in the divine, but instead are the natural byproduct of suppurating tissue. Like the ship in peril, the man's body is unstable and open to dissolution and corruption in the storm-tossed world below the moon. The fundamental corruptibility of his flesh is made manifest in the crab-shaped and worm-ridden ulcer, a condition treated in medical texts under the headings of *De Apostemate*, *De Vermibus in Vulneris*, and *De Cancro*, the last of which was deemed incurable.[29] That Saint Mark could cure this condition with the touch of his ethereal hand proves his divinely appointed power over nature; just as he saved the sailors from shipwreck by preventing the seas and the winds from ripping the ship apart, so too did he save the young man from complete dissolution through his mastery over the human body, purging it of its invaders, closing the aposteme, and making him whole once again.

Our third miracle from the *Legenda Aurea*, attributed to Saint Dominic, the founder of the Orders Preacher canonized by Pope Gregory IX in 1234, synthesizes older *topoi*, such as worms as divine punishment for sin, with newer concerns such as the necessity for the expulsion of worms as a means of cure and as visible proof of a saint's thaumaturgic power. The miracle takes place on the feast day of Saint Dominic in the city of Augusta, where a group of devout women returning from church see another woman busily spinning yarn despite the injunctions against doing so on a holy day of obligation. When confronted with this transgression, the spinning woman responds nastily, saying, "Those of you who are women of the friars," which we might take to mean members of the Dominican Tertiary Orders, "you keep their feasts." Upon uttering these words, the woman's eyes began to swell and pop, and from them issued fetid matter and worms in such a number that eighteen worms were removed from one eye alone.[30] The devout women rushed the impious woman to the church of Saint Dominic; there, she confessed her sins and promised both to keep his feast day holy and to be reverent to his servants—whereupon all of the rotting matter and worms were purged from her eyes and she was completely healed.

The source of illness in this miracle—the worm seed, if you will—is the sin of laboring on a holy day. That the woman was spinning thread and not baking bread or milking a cow is significant; daily chores would always have to be done for the survival of a family, but the production of thread for weaving was lucrative, and not done out of sheer necessity. The woman's sinful nature, evident in her disrespect of the saint and his emissaries as well as her overarching desire for financial gain, was made manifest through the infestation of worms. Because her vision was clouded by greed, the worms hatched from her eyes, a revolting condition that could not easily be hidden; her festering eyes served as an indelible and public sign of her spiritual corruption, condemning her to a life of darkness, away from the healing light of God. For this wormy sinner, the profession of faith and the authoritative sacrament of confession in conjunction with the divine power of Saint Dominic's relics were the sources of spiritual healing; the purification of the woman's soul led to a concomitant purgation of her flesh, thus the worms and other rotting matter were expelled from her darkened eyes, allowing for her illumination.

Worm infestation serves not only a spiritual and didactic purpose in the preceding miracle, but a political and bureaucratic one as well. Saint Dominic lived and died during a period of spiritual renewal in Europe; the thirteenth century saw the establishment of new orders such as the Dominicans, Franciscans, and the Augustinians, which were in theory dedicated to a mendicant life of preaching and service to the laity.[31] The increased role of the laity in spiritual and religious life was reflected in the cult of saints, which saw a surge in regular, mendicant, and lay individuals being promoted for sainthood.[32] Concomitant with this phenomenon, the very process of canonization was also changing; before the thirteenth century, sainthood was proclaimed at the local level. Individuals who were deemed to have lived a holy life and to have performed miracles both in life and death developed cult followings that were sanctioned by the local bishop. Many other saints were martyrs of the early church whose relics had traveled throughout Europe; these saints needed no official approval because of their inherited status. In the thirteenth century, however, the Pope and members of the curia instituted a formal process of canonization through which all candidates for sainthood must pass. This new set of procedures not only signified a shift in power from local bishops to the centralized authority of the Papacy, but also reflected the broader trends toward systematization rooted in the intellectual culture of the medieval university. In the academic milieu, Aristotelian categorical and logical argumentation provided the discursive foundation for the three professions linked to advanced study—law, theology, and medicine—and the authoritative discourse from each of these professions shaped the process of canonization. Where the authority of a local bishop supported by general miracles and the consensus of his community was once enough to proclaim sainthood, the Papacy now demanded a highly detailed dossier containing the life and miracles of a proposed saint supported by legal testimony from supplicants, witnesses and, in some cases, physicians—in other words, proof of the miraculous.

Knowing that their miracles would be passing through the legal process of inquisition, hagiographers used the basic accounts of *miraculés* to sculpt miracles founded in two forms of authority: the unquestionable authority of Christian tradition and the authority of learned medicine and the physician, both of which are reflected in the Saint Dominic's miracle above. Theologically, the miracle reinforces the authority of the Church; the woman is punished for laboring on a Feast Day and is redeemed through the holy sacrament of confession administered by a priest. Furthermore, the *topos* of the worm as divine punishment can be traced back through unquestionable sources such the life of Saint Simeon, the literary tradition surrounding the death of Herod, and the Book of Job. Medically, the woman's worm infestation was a recognizable condition recorded in authoritative medical texts such as the *Viaticum*. Ocular worms were not only tangible and visible, and thus real, but also almost completely incurable; even if the worms could somehow be expelled, the restoration of vision to festering eyes was physically impossible. That such a condition was incurable and beyond spontaneous healing could be subsequently attested to by a physician, thereby proving the miraculous nature of the cure.

This new process for canonization, which demanded legal testimony and medical proof of miraculous healing, facilitated the departure of the snake in the belly *topos* from thirteenth-century miracle tales. In this milieu, the idea of internal infestation by snakes became untenable; poisonous snakes were neither prevalent in the European environment nor prone to being ingested—

snakes of any ilk did not enter into the human body, nor would they survive in the stomach if they did. Furthermore, while snake bites were a topic of concern in medical texts, snake infestation did not exist as a category of illness. Medical authorities were far more concerned with conditions caused by a different legless invader, the worm, which penetrated and thrived within the human microcosm. Learned medical treatises contained sections on worms that infested the eyes, ears, nose, teeth, stomach, intestines, and testes; recognizable and repulsive to a lay audience, rich in theological symbolism of divine punishment and spiritual corruption, the worm functioned as an insidious invader that must be purged for the patient to survive. All of these complexities are woven together in the thirteenth-century miracle of Saint Peter of Verona with which we began. Here, it is not a snake, but a worm that infests the father's stomach and causes his suffering; and it is not a physician—but a saint drawing on the authority of the physician and learned medicine—who administers the purgative of divine grace, thus drawing the worm, two-headed and furry, from the basin of the flesh.

Acknowledgements

I would like to acknowledge the support of Fulbright Scholar Program in both the United States and in Spain, as well as the Huntington Library and Wellcome Unite for the History of Medicine. Additional thanks and loving gratitude go to Brian Walter for all of his wonderful support.

Notes

1. See David Prudlo, *The Martyred Inquisitor: The Life and Cult of Saint Peter of Verona* (Aldershot: Ashgate Press, 2008)
2. "Vita Sancti Petri Martyris de Ordine Fratrum Predicatorum," Acta Sanctorum (*AASS*)
3. This, in and of itself, was miraculous; during the thirteenth and fourteenth centuries, the process of canonization had become so mired in bureaucracy that few potential saints attained the status that their followers had conferred on them informally. See Andre Vauchez, *Sainthood in the Later Middle Ages* (Cambridge: Cambridge University Press, 1997).
4. *AASS*, 4:33, 696.

5. Sherry Reames, *The Legenda Aurea: A Reexamination of its Paradoxical History* (Madison: University of Wisconsin Press, 1985). For a critical edition of the *Legenda*, see Giovanni Paolo Maggioni (Sismel: Florence Italy, 1998).
6. Jacob de Voragine, Legenda Aurea, "The Life of Saint Silvester."
7. The de Voragine's "Life of Saint Philip" includes the story of a serpentine dragon that emerges from a statue of Mars and breathes infected air.
8. Gregory describes the plague as *gravi peste radix infecta est*, and attributes it to an *innumerabilis multitudo serpentem cum magno dracone* in the Tiber River causing "putredine totum aerem corruperunt." Bibliotheca Nacional de España (BN), MS 11544: folio 13v.
9. Flavius Josephus, Antiquities of the Jews, 17:6:5
10. *Ibid.* In Josephus' version, Herod developed a fever that "augmented his pains inwardly" because it gave him a vehement appetite, which in turn caused pain in his ulcerated intestines and especially his colon. Furthermore, "an aqueous and transparent liquor," perhaps humoral phlegm, "also had settled itself about his feet, and a like matter afflicted him at the bottom of his belly"—symptoms that suggest a form of *hydropisim*. "Nay, further, his privy-member was putrefied, and produced worms."
11. For more on the medieval blood libel, see Miri Ruben, *Gentile Tales: The Narrative Assault on Medieval Jews* (New Haven: Yale, 1999).
12. See also the *Book of Judith*, 16:21: Dabit enim ignem et vermes in carnes eorum ut urantur et sentiant usque in sempiternum.
13. See also Frederick Lent, trans., "The Life of St. Simeon Stylites: A Translation of the Syriac Text in Bedjan's Acta Martyrum et Sanctorum, Vol. IV," *Journal of the American Oriental Society*, 35 (1915): 103–198.
14. Jacob de Voragine, *Legenda Aurea*, "Life of Simeon Stylites."
15. Jacob de Voragine, *Legenda Aurea*, "Life of Saint Simeon."
16. Susan B. Mattern, *Galen and the Rhetoric of Healing*, (Baltimore: Johns Hopkins University Press, 2008), Appendix B, Case 300, 197.
17. J. Bondeson, "The Bosom Serpent," *Journal of the Royal Society of Medicine*, 91 (1998): 442.
18. Peter E. Pormann, "The Physician and the Other: Images of the Charlatan in Medieval Islam," *Bulletin of the History of Medicine*, 79:2 (2005): 198.
19. Peter Pormann and Emilie Savage-Smith, *Islamic Medicine* (Edinburgh: Edinburgh University Press, 2007).
20. See Galen, *On The Passions and Errors of the Soul*, trans. Paul W. Harkins.
(Columbus: Ohio State University Press, 1963) and *The Spiritual Physick of Rhazes*, trans. Arthur J. Arberry (London: J Murray, 1950).
21. On lovesickness in Islam, see Michael Dols, Al Majnun: *The Madman in Medieval Islamic Society*. (Oxford: Clarendon Press 1992). For lovesickness in the medieval west, see Mary Frances Wack, "The Liber de heros morbo of Johannes Afflacius and Its Implications for Medieval Love Conventions," *Speculum* 62 (1987): 324–344.
22. For an image of *De Morsu Serpentem* from the *Viaticum*, see ttp://diglib.hab.de/drucke/ma-4f-35/start.htm?image=00799.

23. *Legenda Aurea*, "Saints Cosmas and Damian"
24. *Ibid.*
25. *Legenda Aurea*, "Life of Saint Francis of Assisi"
26. Dangers of retrospective diagnosis aside, worm infestation of the nasal cavities in humans (nasal myiasis) is caused by the primary screw worm and is a problem in regions such as the Indian Subcontinent, although cases have been reported in the United States (Texas), as well.
27. Bibliotheca Nacional (BN), Madrid: *MS 1410*: folio 127b
28. *Legenda Aurea*, "The Life of Saint Mark the Evangelist," BN *MS 4197*, folio 76 r.: "Iuvenis quidam cancri morbo in maxilla, vermibus corrrodentibus, cruciatus…"
29. Luke DeMaitre, "Medieval Notions of Cancer: Malignancy and Metaphor," *Bulletin of the History of Medicine* 72 (1978): 609–637.
30. Maggione, *Legenda Aurea*, 741: *Statimque eius oculis in tumorem cum pruritu converses ex ipsis ceperunt vermes protinus scaturire ita quod vicinia quedam XVIII vermes eduxit ex eius oculis in instanti.*
31. Lester K. Little, *Religious Poverty and the Profit Economy in Medieval Europe* (Ithaca: Cornell University Press, 1983)
32. Andre Vauchez, *Sainthood in the Later Middle Ages* (Cambridge: Cambridge University Press, 1997)

READING THE WORMY CORPUS: AMBIGUITY AND DISCERNMENT IN THE LIVES OF MEDIEVAL SAINTS

Alison More

Tensions between body and spirit have created a quasi-dualistic antagonism in Christian thought. Theologians have gone to great lengths to emphasise the importance of transcending the physical realm; however, as embodied creatures, humans need the physical world to convey and understand the language of the symbolic which seldom has a meaning that is both lucid and unequivocal. Animals which are used symbolically both convey and are part of an elaborate literary code.[1] One such symbolic animal is the worm: though at first glance the worm is symbolic of death and decay, burrowing deeper reveals that it is also a symbol of purity and resurrection. Repeatedly, the connection between the physical and the textual worm signifies both life and death and is connected with both the demonic and the divine. This article first discusses medieval understanding of the worm, then explores the ways in which reading the wormy corpus is an exercise in discernment.

The Worm

In his *De doctrina Christiana,* Augustine commented on the need for an encyclopaedia that would explain the workings of the natural world. Throughout the Middle Ages, a number of thinkers rose to this challenge and wrote works exploring the different aspects of the natural world.[2] The influential encyclopaedist Isidore of Seville speaks of many different kinds of worms: worms of the air (spiders); worms of the water (leeches); worms of the earth (scorpions, snails, and beetles); clothing worms (moths); wood worms (termites); and leaf worms (caterpillars). Worms of the flesh (*vermes carnium*) fed on human remains, specifically intestines and skin. To Isidore, the worm was essentially unclean, and had some traits in common with its larger cousin the snake.[3] He perceived that the one redeeming feature of worms was that they did not engage in sexual reproduction.

Medieval understanding of the worm was exceedingly complex. The encyclopaediae from the high Middle Ages largely echo Isidore's writings, but demonstrate that Isidore was far from capturing the authoritative view on worminess. A thirteenth-century example of such work, Thomas de Cantimpré's *De naturis rerum,* contains a substantial section dealing with worms. While Thomas' other writings – notably his *vita* of Christina (a holy woman from St. Trond) and his *Bonum universale de apibus*, a text chronicling the more fantastic accounts of religious experience in Liège – demonstrate a certain incredulity, *De naturis rerum* is somewhat conventional.[4] In this work, Thomas draws upon the writings of Isidore, Ambrose, and his mentor Jacques de Vitry. Instead of finding synthesis, Thomas adds another dimension to Isidore's complex category of worminess by insisting the Toad – another venomous creature of dubious character – was a type of worm.[5]

As in other aspects of the medieval understanding of the natural world, the depiction of worms found in Isidore's writings was dominant. Worms were essentially pestilential beings who fed off of the life-substance of both human kind and the natural world. However, another tradition presented a more positive spin on the worm. Many medieval Bestiaries were influenced by the *Physiologus,* an anonymous Greek text detailing the physiology of animals.[6] Numerous Latin translations of the *Physiologus* appeared throughout the Middle Ages. Both its availability and proximity to the learning of the classical world ensured that – despite its often tenuous connections with the natural world – the knowledge contained in the *Physiologus* formed the basis for the physiological and symbolic interpretations of the animal kingdom which are found in many bestiaries.

The *Physiologus* added a further dimension to the already-complex medieval concept of worms. Instead of filth and decay, the humble worm finds certain redemption in this text: the worm is not related to the snake but the phoenix. According to legend, once the phoenix has lived for five hundred years, it voluntarily sacrifices itself. Working with a temple priest, the phoenix anoints itself and submits to a sacrificial fire. The next day, the priest finds a small worm in the ashes – which transforms itself into an eagle by the third day.[7] The phoenix had been a symbol of death and resurrection in the classical world, and was often portrayed on funerary urns. In the Christian world-view, both the phoenix's willingness to sacrifice itself and its resurrection after three days created a logical association between the phoenix and Christ.[8] The *Physiologus* reads:

> The phoenix represents the person of the Savior since, descending from the heavens, he left his two wings full of good odors (that is, his best words) so that we, holding

> forth the labours of our hand, might return the pleasant spiritual odor to him in good works.[9]

Christianity's often disparaging attitude towards sexuality combined with its emphasis on humility to underline a puzzling array of Christomimetic worms. The worm's ability to reproduce without recourse to sexual intercourse invited an association between worms and sexual purity. Unlike other animals, the worm was never sullied by carnal lust, but was generated seemingly spontaneously from wood, flesh, meat, or any decaying matter.[10] Thomas de Cantimpré and others took this as a manifestation of the Christological mystery in the natural world: the spontaneous—even miraculous—generation of the worm was, in some way, analogous to the virgin birth.

Of course, it had not escaped notice that the non-seminal origins of the worm occurred in human corpses, decaying filth and even excrement. Because of this Albertus Magnus perceived them as having many similarities to demons. He pointed out that, like worms, demons originated in filth and could only sustain their forms for a short time. Criticism was not unique to the Christian theological tradition.[11] Instead, Jewish writers and thinkers such as Jacob B. Reuben were all too ready to formulate satirical parodies and scathing intellectual attacks on the peculiar Christian deity who came into the world through filth and squalor, and who was likened to a worm.[12]

Medieval thinkers viewed all aspects of the physical world as having many layers of meaning. A literal interpretation of anything would merely capture a single expression of its essence. In his influential *De doctrina Christiana* Augustine addressed the multiple layers of meaning:

> The variation takes two forms. In this way, one thing signifies another thing and, moreover, another. This is done either in such a way that the second thing signified is different from or contrary to the first ... This is the situation where the Lion is used to signify Christ when it says 'the lion of Judah has prevailed,' but also signifies the devil when it is written: 'your adversary, the devil, goeth as a roaring lion seeking whom he may devour.'[13]

Here, the symbol of the lion does not change: it simply embodies two seemingly contradictory realities. The same is true of the worm: on the one hand, it is the embodiment of purity and humility; on the other, death, and decay. Despite a link between the worm and Christ, worms were most often associated with the female body: in its positive guise, the worm was associated with the paragon of Christian femininity, the blessed Virgin Mary; in its negative guise, the worm was associated with decay of the flesh. This was not only due to the innate connections believed to exist between women and the physical, but also drew upon the theme of decay. Decay of the flesh was both closely connected with feminine fertility, and more importantly, a consequence of the fall and a link with Eve. [14]

Divine Worms

Rather than a single puzzling instance of wormy divinity, the analogy between Christ and the worm is expounded upon in a number of medieval texts. Patristic exegetes making this connection often draw upon Psalm 21:7 which reads, "I am a worm and not a man," which is often interpreted as a reference to Christ. A Bestiary currently in the Vatican manuscript collection reads:

> Quie sponte intravit super aram, id est, super crucem ubi ipse sacerdos optulit q seipsum patri pro nobis. Et altera die inventus est qui vermis in sepulcro. Unde dicitur, Ego sum vermis et non homo. Tertia die reformatur in pristina forma, id est, resurrexit a morte et sic xl die pariter ad primum suum in celis.[15]

Although the wormy Christ is far from a mainstream figure in medieval texts, the exegetical tradition which draws this parallel was both endorsed and expounded on by figures such as Jerome and Augustine. [16]

Diabolical Worms

Worminess had another, contradictory meaning: the embodiment of decay and punishment for sin. The seemingly innate negative associations of the worm are reinforced in scripture. The book of Isaiah reads "And they will go out and look upon the dead bodies of those who rebelled against me; their worm will not die, nor will their fire be quenched, and they will be loathsome to all mankind."[17] This verse has a later gospel parallel which speaks of the punishment of sinners, saying "...It is better for you to enter the kingdom of God with one eye than to have two eyes and be thrown into hell, where 'the worm does not die, and the fire is not quenched.'"[18] The book of Sirach associates worms with Hell, saying that "what awaits man is worms."[19] Similarly, the book of Judith associates worms with punishment of the unrighteous saying "...in the day of judgment he will punish them: He will send fire and worms into their flesh, and they shall burn and suffer forever."[20] These scripture passages have influenced numerous portrayals of Hell, in which the damned are portrayed as food for worms.[21] Texts recounting an apocalypse and other visionary literature typically portray worms and other beasts of torment feasting on sinners.[22]

The ways in which the worm attacked the physical body were evident on a literal level. Medieval texts attribute allegorical and anagogical meanings to the worm by discussing its role in gnawing on the souls suffering in the afterlife. Augustine writes that after the final judgement the damned will be re-united with their bodies where they will both suffer physical pains and their souls will be "gnawed by the worm of regret."[23] This persisted through the high Middle Ages: Hildegard of Bingen frequently describes sin as "putrefying matter which gives rise to pus and worms";[24] and twelfth-century text known as the 'Visions of Tondal' describes worms devouring sinners in the afterlife.[25] Thomas Aquinas' *Summa Theologica*:

> The worm ascribed to the damned must be understood not of a corporeal but a spiritual nature: and this is the remorse of conscience which is called a worm because it originates from the corruption of sin, and the torments of the soul as a corporeal worm is born of corruption and torments by gnawing.[26]

Aquinas' passage and others that are comparable indicate that the worm had not lost its associations with sin and decay by the high Middle Ages. Instead, the association between worms and decay were continually cited as evidence in support of the Virgin's Assumption: as Mary's body was the source of Christ's humanity, it was inconceivable that it could be consigned to the posthumous fate of being devoured by worms.[27]

Worminess and Discernment of Spirits

It is one thing to point out that the symbol of the worm was multifaceted; it is another to examine its use in medieval texts. Worms of uncertain soteriological affiliation crawl through the pages of medieval texts and are depicted in art. In particular, instances of both the worm's varied and oscillating use and its alternately divine and diabolic associations are depicted in hagiography.

In medieval thought, very few factors distinguished liminal experiences, those extraordinary moments that push the boundaries of consciousness, that were divine from those diabolic in origin. As the two types of ecstatic experience were often disturbingly similar in appearance, it was increasingly important that these experiences were approved by a confessor and recorded in a way that could leave no doubts concerning the orthodoxy of the subject. While there are instances of worminess linked with possession, there are also instances where wormy infestations seem to have a salvific dimension. Depicting both the process of discernment and the experience often involves a symbolic language being written on the canvas of the body.

Despite the often ambiguous effects of mystical experience, there were some unmistakeable signs of whether the individual in question was afflicted by divine or diabolic spirits. As Caesarius of Heisterbach recounts:

> Anyone who doubts that demons exist should take note of their energumens, for when the devil speaks through their mouths and rages so cruelly in their bodies, he shows them a manifest signs of his presence.[28]

Still, discernment was far from straightforward: elsewhere in his chapter on demons, Caesarius recounts tales of them appearing in Church or being familiar with holy books.[29] This was not unique to Caesarius: the visions of Ermine of Reims contain numerous tales of demons appearing in the guise of numerous, angels and saints, her deceased husband, or her confessor. Nevertheless, Ermine remains steadfast in her ability to discern.[30]

Medieval texts reveal considerable overlap between demons and the gruesome images of worms feasting on the bodies of the damned. These texts spoke in a language of symbols, which dictated that the physical forms of demons were gruesome, representative of their true nature. Textual demons took the form of toads, serpents, worms (often covered with black hair), or black imps.[31] Similarly, demons were often portrayed in animal, in particular, cats, leeches, and worms.[32] The ambiguity of the worm is not unique: instead, the language of symbols was particularly ambiguous where animals were concerned. Although numerous texts and pictures use animal symbolism, explicit statements about its meaning are uncommon.[33]

Barbara Newman states that possession can be defined as "an indwelling malevolent spirit."[34] While medieval texts indicate that possession was believed to be physical in nature, there were a variety of opinions regarding the ways in which demons gained access to the body: Caesarius of Heisterbach recounts an incident in which a demon attempts to hide itself in milk;[35] while the *vita* of Ida of Nivelles records that the saint invited a demon to enter her body to relieve the torment of her possessed sister. The physical entry of demons was complemented by tales of physical exit: exorcisms routinely describe demons exiting the body through orifices such as the mouth or anus.[36]

Thomas de Cantimpré's *Bonum Universale de Apibus* recounts the exorcism of a young girl who opened herself to demonic possession while dancing on a Sunday. By uttering the name of Christ, a young boy compels the demon to make itself known. The demon, which is powerless to resist the divine name, appears as a lump beneath the girl's navel. By repeatedly making the sign of the cross on the girl's flesh, her exorcist compels the demon to move up to her mouth. Here, the demon exits through the girl's mouth in the guise of a black and hairy worm.[37]

As demons were thought to have gained physical entrance to the body of one possessed, possession was thought of as a physical illness, and it was therefore unsurprising that the presence of demons would make itself known through physical symptoms, which were often similar to the demoniac described in Mark's gospel: wandering among tombs, extraordinary strength, continual wailing, and beating himself with stones.[38]

Physical changes such as miraculous knowledge of languages, extraordinary strength, and loss of bodily control were not only signs of an indwelling demon, but could also signify a mystical ecstasy. Thomas de Cantimpré's *life* of Christina of St. Trond presents the tale of a woman whose disturbing strangeness seems to provide both the basis for her sanctity and grounds for calling it into suspicion. Christina was born in a village outside Liège in modern-day Belgium, and orphaned at a young age. Near the beginning of the *vita,* Thomas recounts that Christina died. During her funeral mass Christina's body floated to the rafters of the church during the *Agnus Dei* where it remained until the priest commanded Christina to return to earth. Christina then returned home with her sisters where she was 'reinvigorated by food.'

Christina's account of her experience was no less astonishing than the events at her funeral. She told her anxious spiritual friends while her body lay dead, her soul was led through purgatory, where she witnessed the torments "so many and so great that no tongue is adequate to tell of them." She witnessed similar pains in Hell. After this, she was sorrowfully brought to Heaven, where she was offered two choices "either to remain with [the Lord] … or to return to the body and suffer the sufferings of an immortal soul in a mortal body."[39]

Christina chose the path of selfless suffering, and her use of physical asceticism was nothing less than dramatic. Indeed, Thomas recounts that Christina crawled into ovens and was hung from the gallows by her own hand.[40] Christina ate such filth that it often seemed that she was eating toads which, as was mentioned above, were a type of worm in Thomas' mind. At the same time Christina manifested extraordinary strength, the ability to prophecy, and an aversion to human society – all of which were as characteristic of *possessae* and *obsessae* as of holy women.[41] Indeed, throughout the *vita Christinae* Thomas acknowledges that Christina's friends and relatives frequently attributed her behaviour to possession.

Instead of being unique to Christina, this difficulty in discerning mystical ecstasy and saintliness from demonic possession was common throughout Europe. For example, when speaking of Margaret of Cortona's mystical ecstasy, her hagiographer Giunta Bevegnati, recounts an incident which blurs the boundaries between possession and mystical ecstasy:

> For they saw Margaret, not beside the Cross, but having endured bitter sorrows as if she had been placed on the Cross. In which, many wondrous signs of grief were apparent that we believed that she were nearly placed in the hand of death. She would gnash with her teeth in grief, be twisted like a worm and be discoloured to the likeness of ashes, she would lose her heartbeat, she would lose her voice, she would become completely rigid, and thus the upper parts of her throat were made hoarse so that she could scarcely be understood when she returned to her senses.[42]

Psychological and physiological disturbances were a common sign of possession. Becoming completely removed from the sensible world – being taken *ex stasis* – was associated with both instances of divine and demonic. A similar tendency is noted in the auto-hagiography of Angela of Folingo. Angela of Foligno's confessor, Brother 'A.' recalls one of her earliest mystical experiences in which she began shouting words which became almost intelligible outside the basilica in Assisi.[43] It was only after intimate discussions and careful discernment that he came to believe in her sanctity. Rather than hiding the extraordinary aspects of her behaviour, Brother A. uses them in promoting Angela as a model of holiness.

Rather than being exceptional, these women illustrate a norm. Similar slippage of categories can be seen among medieval holy women, regardless of geographical location or religious affiliation.[44] The majority of texts which convey encounters with either divine or diabolic spirits do so in a way which, at first glance, appears wholly ambiguous. Whether her indwelling spirits are divine or demonic, both the role and appearance of a *possessa* are similar and the process of discernment difficult and subtle.

As possession was physical, discernment of spirits could often be accomplished through discernment of bodies; however, at times it was less clear. A number of divine experiences portray physicality in ways which appear unholy. Alice of Schaerbeek frequently loses body parts to leprosy while the flesh of her later counterpart Lidwina of Schiedam crawls with worms.[45]

All too often, leprosy was seen as punishment for sin; decaying flesh, particularly flesh that is being devoured by worms evokes overtones of the traditional punishment for sin. At the same time, physical illness could be seen as salvific. In particular, the leper was a powerful reminder of the physical nature of suffering, particularly vicarious suffering. In many instances, such as the cases of Lidwina and Alice, the torments of the flesh were used to enhance the portrayal of holiness and were linked with their individual spiritual journeys.[46]

Alice was a Cistercian nun, and a member of the community of Schaerbeek. On Good Friday, Alice received a vision of Christ crucified: his body weakened, pierced, and drenched in his own blood. Christ begged Alice to remember what he had endured and asked her to imitate him. Christ also gave Alice a gold cross, which she interpreted as a call to share in his sufferings.[47] Alice's opportunity to participate in redemptive suffering came in the form of leprosy. As her sisters were fearful of catching the illness, they required that she live in a separate hut on the convent grounds and to stand apart from the community during liturgical services. Instead of viewing her physical pain and isolation as a punishment, Christ assured Alice that her leprosy was a gift, a sign of perfect love. Christ had allowed this suffering so that Alice might have leisure to be with her divine bridegroom in the "bridal-chamber of her mind."[48] At the same time, Alice dedicated her pain to making satisfaction for the souls in purgatory.

While Alice had the official blessing of a religious community, Lidwina, as a *soror in mundo*, was regarded with some suspicion.[49] Lidwina's own mother doubted her sanctity before it was confirmed miraculously. As a young woman, Lidwina suffered a fall which injured her ankle. Rather than an isolated incident, this unfortunate moment began a series of physical ailments, each more dreadful than the last. From the moment of her fall, Lidwina suffered ailments ranging from loss of control of her limbs to frequent fainting spells. Later, Lidwina's hagiographer recounts details which seem to suggest her flesh was rotting while still attached to her body. All too often leprosy was seen as punishment for sin. Decaying flesh, particularly flesh that is being devoured by worms evokes overtones of the traditional punishment for sin.

Long before her hagiographer describes the horrific effects that Lidwina's suffering inscribed on her body, he is careful to point out that she endured these pains to make satisfaction for the suffering souls in purgatory. He assures his readers that both the pains and the form they took were pleasing to God.[50] While the audience of the *vita* would undoubtedly shudder at the description of the worms that crawled from one of the three orifices in Lidwina's stomach. The *vita* recounts that these worms were grey in colour, about the size of a human finger, and covered with black hair. Doctors who examined Lidwina were astonished, less by the beasts gnawing her innards than the fact that her disintegrating flesh emitted a sweet odour.[51]

Although Lidwina cured many who were suffering, she made no effort to relieve her own pain.[52] Her hagiographer recounts an incident in which Lidwina both made a sacramental confession in place of an evil man and carried out punishment on his behalf. Similarly, Lidwina voluntarily took on the pains of a woman suffering from toothache, The name 'Lidwina' is form of 'Lidje', a derivative of the Dutch 'lijden' meaning 'suffering.'[53] The *vita* portrays Lidwina as the embodiment of suffering. Vicarious suffering is the sign and instrument of perfect love. Through her own suffering and the suffering of others, Lidwina demonstrated that she not only adhered to, but also internalised the imitation of Christ. Because of her sufferings Lidwina was able to say "I bear the marks of Christ in my body" and "I no longer live, but Christ lives in me."[54]

Despite their fleshly torments, both Alice and Lidwina can be placed in the tradition of feminine sanctity. Both were pious from their childhood, both persevered in virtue, and the hagiographers of both women were able to portray their afflictions in a pious light. Worminess was torment, but not all torment was demonic. The multivalent symbol of the worm had significance on many levels. While Thomas de Cantimpré's black and hairy worm inside the bowels of a young girl indicated the presence of a demon, the worms that gnawed on Lidwina's bowels were salvific. Worms were neither good nor evil, neither divine nor diabolical. Like all of creation they had the potential for both. The ambivalence of the worm is a continuous wriggling strand through medieval vernacular theology.

Like Lidwina, the worm is a symbol which is neither one thing nor the other – instead both have become living embodiments of conflicting realities which can only be relegated to the realm of ambiguity. Lidwina's suffering was evidence of both her fallen nature and her desire to become divine. Like the worms which consumed her flesh, there were many layers of meaning in both Lidwina's existence and her torment. In an instance of Christomimesis, Lidwina did not endeavour to escape her suffering, but allowed herself to be transformed through it. In their wholly liminal and seemingly ambiguous positions, both Lidwina and the worm are being presented in a Christomimetic form.

It is important to keep in mind that Christ himself is an astonishing coincidence of opposites. The later medieval emphasis on Christ's humanity led to an increase in the attention given to his passion and heightened the sense of paradox. Though innocent, Christ had effected satisfaction for the debt incurred by human sin; men and women of the high Middle Ages sought to imitate this aspect of Christ's earthly existence through castigating their own flesh.[55] Through his sacrifice, Christ had achieved the ultimate triumph: conquering sin and death. Human beings, by nature, are fundamentally incapable of such a victory; however, saints like Lidwina were able to participate in the mystery of redemption. Symbols such as the worm – a wholly ambiguous creature in the medieval symbolic lexicon – were able to transmit a small part of that reality.

Notes

1. Jan M. Ziolkowski, "Literary Genre and Animal Symbolism," in *Animals and the Symbolic in Medieval Literature*, ed. L. A. J. R. Houwen (Groningen: Egbert Forsten,

1997), 1–23. Cf. Alexandra Cuffel, *Gendering Disgust in Medieval Religious Polemic* (Notre Dame: University of Notre Dame Press, 2007), 198.

2. Lynn Thorndike, *History of Magic and Experimental Science*, volume 4 (New York: Colombia University Press, 2003), 374–78.
3. Isidore of Seville, *Etymologies* (Cambridge: Cambridge University Press, 2006), bk 12, 5:1–18.
4. Thorndike, *History of Magic*, 376.
5. Thorndike, *History of Magic*, 385. Cf. William Monter, "Toads and the Eucharist: The Male Witches of Normandy, 1564–1660," *French Historical Studies* 20 (1997): 563–95.
6. Debra Hassig, *Medieval Bestiaries: Text, Image, Ideology* (Cambridge: Cambridge University Press, 1995), 72–75.
7. *Physiologus*, trans. Michael J. Curley (Austin: University of Texas Press, 1979), 73.
8. Hassig, *Medieval Bestiaries*, 72–75.
9. *Physiologus*, 73.
10. Cf. Caroline Walker Bynum, *The Resurrection of the Body in Western Christianity, 200–1336* (New York: Columbia University Press, 1995), 24–26; 29–32.
11. Thorndike, *History of Magic*, 484–86.
12. Cuffel, *Gendering Disgust*, 22, 133.
13. Augustine, *De doctrina Christiana*, PL 34, 3.25.6.
14. Bynum, *Resurrection*, 221.
15. Vatican Bestiary, f. 29v. Cited in Hassig, "Medieval Bestiaries," 75.
16. On the worm in this psalm as a figure of Christ, see Augustine, *Ennarationes in psalmos*, PL 36–37: 168, and Jerome, Brevarium in psalmos PL 26 932D–933A.
17. Isaiah 66: 24.
18. Mark 9: 47–48.
19. Sirach 7:17.
20. Judith 16:17.
21. Clifford Davidson, "The Fate of the Damned in English Art and Drama," in *The Iconography of Hell* (Kalamazoo: Medieval Institute Publications) 42 and 52; Emile Mâle, *Religious Art in France in the Thirteenth Century*, trans. Dora Nussey (London: J.M. Dent, 1913), 380–81.
22. Jacques Voisnet, *Bêtes et hommes dans le monde medieval* (Turnhout: Brepols, 2000), 183, 187–89; Martha Himmelfarb, *Tours of Hell: An Apocalyptic Form in Christian and Jewish Literature* (Philadelphia: University of Pennsylvania Press, 1983), 116–19.
23. Bynum, *Resurrection*, 95–97.
24. Bynum, *Resurrection*, 159, n. 6.
25. Bynum, *Resurrection*, 159.
26. *Summa Theologica*, q. 97, art. 3, supplement, pt. 3.
27. Pseudo-Augustine, "Ad interrogate: De Assumptione Beatae Mariae Virginis," PL 60 col. 1146, ch. 6. Many medieval thinkers affirmed this in relation to Christ's body (Bynum, *Resurrection*, 170) .

28. Caesarius of Heisterbach, *Dialogus Miraculorum*, ed. Joseph Strange, 2 vols. (Cologne: Heberle 1851). (Hereafter 'DM'. 5.12.
29. DM c. 5.
30. *Entre Dieu et Satan: Les Visions d'Ermine de Reims (+1396)* trans. Claude Arnaud-Gillet (Florence: SISMEL Edizioni del Galluzzo, 1997), 55–70. Similarly, Catherine of Genoa sees Satan in both the guise of the blessed Virgin and Christ. See Catherine of Genoa, *Seven Spiritual Weapons*, tr. Hugh Feiss, (Toronto: Peregrina Press, 1998), c. 7.
31. Nancy Caciola, *Discerning Spirits: Demonic and Divine Possession in the Middle Ages* (New York: Cornell University Press, 2003), 41, esp. n. 26. Cf. Barbara D. Palmer, "The Inhabitants of Hell: Devils," in *The Iconography of Hell* (Kalamazoo: Medieval Institute Publications, 1992), 20–40 (21–22).
32. This association existed as early as patristic times; cf. David Brakke, *Demons and the Making of the Monk* (Cambridge: Harvard University Press, 2006), 107–08.
33. Ziolkowski, "Literary Genre," 5.
34. Barbara Newman, "Possessed by the Spirit: Devout Women, Demoniacs, and the Apostolic Life in the Thirteenth Century," *Speculum* 73 (1998): 733-70 (736).
35. DM passim.
36. Caciola, *Discerning Spirits*, 200–03.
37. Thomas de Cantimpré, *Bonum Universale de Apibus*, (Douai, 1627), 2.36.4, 386–87.
38. Mark 5: 3–9. Cf. Newman, "Possessed by the Spirit," 733–70.
39. VCM bk. 1, c. 2. par. 6, 252.
40. VCM bk. 1, c. 1, par. 11, 652 (fire); ibid, bk. 1, c. 1, par. 13, 652 (gallows).
41. Newman, "Possessed by the Spirit," 736.
42. Giunta Bevegnati, *Legenda de vita et miraculis beatae Margaritae de Cortona*, ed. F. Iozzelli (Rome, 1997), c. 5, par. 3.
43. Angela of Foligno, "Memoriale," in *Angela of Foligno, Complete Works*, trans. Paul LaChance (NY: Paulist Press, 1993), 141–42.
44. Caciola, *Discerning Spirits*, 129–75.
45. *De B. Aleyde Scharembeka, Sanctimoniali Ordinis Cisterciensis, Camerae Iuxta Bruxellam*, in *Acta Sanctorum*. June, II, 477–81. (Hereafter VAS)
46. For further exploration of this in the life of Alice, see Martin Cawley, "The Life of Alice and the Silver Age of Villers," *Cistercian Studies Quarterly* 31 (1996): 51–74.
47. VAS c. 2, par. 8–9, 479.
48. VAS c. 2, par. 9, 479.
49. These suspicions gave rise to rumours that Lidwina's swollen abdomen was not a result of her illness, but because she had been made pregnant by the local priest.
50. *De s. Lidwige sive Lidwina virgine schidami in hollandia, Prior vita, Acta Sanctorum*, April, II, 270–302. (Hereafter, VLS).
51. VLS c. 1, par. 10, 275.
52. VLS c. 8.
53. *Encyclopedia of Disability*, v. 2, ed. Gary Andrecht (New York: Sage Publications, 2005), s.v. Lidwina, 136.

54. Cf. Gal 6.7
55. Esther Cohen, "The Animated Pain of the Body." *The American Historical Review* 105 (2000): 36–68.

Some Considerations Regarding the Origin and Functions of Parasites among Two *Mbya* Communities in Misiones, Argentina

Marta Crivos, María Rosa Martínez, Carolina Remorini, and Anahí Sy

Ethnographical and ethnological studies such as the pioneering work of Levi-Strauss and Evans-Pritchard suggest that in the religious and traditional belief systems of indigenous populations, certain people, animals and spirits, are said to have the power to injure individuals and/or the community as a whole.[1] Physical injuries, including sickness and death, are often inflicted by entities in response to perceived transgressions of fundamental laws and social boundaries that regulate relationships between the natural and supernatural realms. In fact, the violation of taboos regulating the interaction between natural and supernatural entities appears to be one of the main causes of illness and death in many indigenous communities. In these cultures, the human and natural environments expand beyond their empirical or biological features to encompass a variety of entities that can transgress permeable boundaries—including those of the human body—at will. The study of indigenous ideas about and representations of illness, especially when caused by supernatural entities, provides a window onto indigenous perceptions of the human body, human relationships, and the interactions that exist between human and non-human entities that live in and share a particular environment. The focus of this study will be the relationship between worms—natural and supernatural, beneficial and harmful—and the people of the *Mbya Guarani* communities of the Argentinean Northeastern Rainforest.

In their everyday activities, members of *Mbya Guarani* communities categorize and utilize their natural environment—the sub-tropical rainforest—and give different values (positive, negative, or neutral) to different components of that environment. The relationships between the *Mbya Guarani* and the non-human entities of the forest that sustains them are surprisingly elastic; some of the most complex and ambivalent relationships are those between *Mbya Guarani* and worms. In many cultures, the penetration of the body by worms is a cause of both fear and disgust. In *Mbya Guarani* culture, however, certain types of worms are seen as completely natural, and their presence and activity in the environment and the human body are thought to be not only harmless, but also beneficial. For example, the worms that grow in the marrow of the *Pindo* palm tree (*Syagrus romanzoffiana*) are ascribed a positive value because they serve as a delicious and nourishing source of protein. There are also worms that are born with and within a person, and live in the human body as their macrocosmic environment; these worms are seen as a natural part of a person's body and not as invaders or as pathogens. Worms from the soil and water that enter the body unbidden, however, can become pathogenic and provoke illness, and are ascribed a negative value.

For the *Mbya Guarani*, then, the presence of worms or "bugs" in the body does not always imply a state of illness. Because of the complex nature of *Mbyan* conceptions of the human body and its connections to the natural and supernatural worlds, the relationship between worms and humans is not always parasitical (beneficial for one species at the expense of the other) but instead is often symbiotic (beneficial to both species). This reciprocal relationship, however, may become unbalanced if the worms' habitual behaviors become pathogenic thereby causing damage to the host's health. For the *Mbya Guarani*, worms not only cause damage but also are the result of "damage" (*paye*), which in this culture is closely allied with maleficence. *Paye* is the negative action of one human being who wants to inflict evil on another human being, which often results in either sickness or death. One of the deadliest forms of *paye* materializes in the shape of a worm that invades and attacks the body, consuming internal organs until the afflicted individual finally succumbs. *Paye* worm infestation is considered by far to be the most frightening; unlike natural worms that serve as food or are born within the body, *paye* are imbued with supernatural power. This kind of situation is the result of an interpersonal conflict between members of the community, which can be due to different reasons. These worms are feared, as they are the expresion of evil, of a conflict.

From the perspective of modern Western biomedicine, the presence of worms in the human body—no matter their nature or source—is always considered to be negative. Health policies set forth by medical and governmental agencies, therefore, focus on the complete elimination of worms by means of medical, pharmacological, hygienic, and environmental intervention. Recent research done on intestinal parasitoses among indigenous people in the Amazon confirms that parasitic worms are a medical problem requiring the attention of government officials, medical authorities, and environmental experts. In the Argentinean rainforest, the combination of clayey, damp soil containing a high concentration of organic materials together with a climate characterized by sustained high temperatures provides a suitable environment for the development, spread and chronicity of enteroparasitosis. Despite the fact that enteroparasitosis are endemic in this region, these pathologies tend to be underestimated by both the general population and the local sanitary staff because they are often asymptomatic. Undetected, parasitic infections wreak havoc within the body and are responsible for increased morbidity, especially when coupled with malnutrition. The integral relationship between geohelminth infections, malnutrition and anemia has been demonstrated by several authors, including Navone et al in their 2006 article, "Parasitosis intestinales en poblaciones Mbyá-Guaraní de la Provincia de Misiones, Argentina: aspectos epidemiológicos y nutricionales."[2]

Our research reveals that while the *Mbya* and the biomedical sector converge in their belief that many types of worms are detrimental to the human body, they diverge in their ideas regarding the origin, mechanisms of pathogenic action, and suitable therapeutic actions to be taken in the treatment of parasitic infection. The growing presence of modern biomedicine in Mbya culture as well as the increase in the number of individuals from these communities who train in sanitary institutions has led to a significant transformation in local conceptions and practices regarding parasites and parasitoses. This new knowledge has been integrated into traditional belief systems, resulting in a changed perception among the *Mbya* of the risks associated to the presence of worms. Our paper analyzes Mbya beliefs and practices concerning the existence of parasites in humans, especially the representations of their origin and behavior in the human body, as well as on practices associated with the diagnosis, prevention and therapy of parasitoses. The information included in this paper is the result of ethnographic and interdisciplinary research on human parasitology carried out between 1999 and 2008 in two Mbya Guarani communities in the Province of Misiones, Argentina. This research was developed by a team of ethnographers, biologists, parasitologists, ethnobotanists and biological anthropologists belonging to the Facultad de Ciencias Naturales at Universidad Nacional de La Plata[3].

The Mbya Guarani Communities

The *Mbya*, together with the *Kayova*, the *Ñandeva* and the *Ava Guarani* or "*Chiriguanos*," are the *Guarani* groups with the largest number of members. They speak the languages belonging to the *Tupí-Guaraní* linguistic family. According to recent estimations[4] the total number of *Mbya* in Brazil, Paraguay and Argentina adds up to about 19,200 individuals. According to the 2005 ECPI (Complementary Survey of Indigenous Peoples), there exist in Argentina around 3,975 people who identify themselves as belonging to the *Mbya Guarani* people in Misiones province.[5] The Mbya presence in Misiones dates back to the end of the nineteenth or beginning of the twentieth century[6] when they begin migrating outward from the mid-southern region of the forest in modern day Paraguay.[7]

The *Mbya* inhabit the Paranaense rainforest, one of the most complex and biologically diverse ecosystems in South America. The Paranaense forest encompasses all of Misiones province and extends northeast to the province of Corrientes in Argentina, and continues eastward into Paraguay and south to Brazil. Over recent generations, the natural resources of this ecosystem have been seriously depleted and, as a result, the very boundaries of the forest have contracted until only a very small amount of its original territory remains. The contraction of the forest is the consequence of several factors, including: the exploitation of timber resources in the rainforest; the substitution of native species for exotic tree species; the building of hydroelectric dams; and the advent of agricultural colonization and the growth of lucrative crops for the export market, such as tea, yerba mate, tobacco and tung tree.[8] Also contributing to the transformation of the rainforest is the sustained use of its natural resources through hunting and fishing, gathering, and "slash-and burn" agriculture —all of which are the subsistence strategies of aboriginal groups in the area.[9]

Mbyan culture has been, until recently, characterized by their constant spatial mobility as they exploit forest resources and search for areas having favorable conditions for *Mbya reko*, that is, the *Mbya* way of life. Periodical movement within the ecosystem's limits has allowed them to recover and reuse previously occupied spaces. At present, the *Mbya* population tends to form small, relatively stable settlements, although some individuals or small family units still move regularly. The more sedentary lifestyle embraced by the *Mbya* in recent years is the result of several factors, the central one of which is the shortage of available land having the conditions necessary for their survival. Much of the land in the Paranaense has been claimed as private property or has been incorporated into the state national parks or designated as natural reserves, all of which either limit or ban the settlement and use of that space. Another factor favoring sedentarization among the *Mbya* is provision of subsidized housing by the government and ENDEPA (National Aboriginal Pastoral Team).[10]

The *Ka'aguy Poty* and *Yvy Pyta*, the two *Mbya* communities that are the focus of our research are settled in *Valle del Arroyo Kuña-Pirú*, in the middle of Misiones province, between the departments of *Cainguas* and *Libertador General San Martín*. These communities are connected with neighboring settlements including *Aristóbulo del Valle, Jardín América,* and *Ruiz de Montoya* as well as with other *Mbya* communities on the basin of *Cuña Piru I* and *II* streams, by Provincial Route N° 7 which slices across the northern part of the Reservation (see Figure 1: Map). Within the *Ka'aguy Poty* and *Yvy Pyta* communities, the *Mbya* language is spoken, and most adults and school children also speak Spanish. As a result of the migration and settlement of communities near convergence of the borders of Argentina, Brazil, and Paraguay, Portuguese and *Yopará* (Paraguayan *Guarani*) are also spoken. According to a census conducted by our team in May of 2003, the population of both communities totals 280 people. From a demographic point of view, it is a "young" population, as the greatest number of individuals is between 0 and 14 years of age (54%). Percentages decrease with an increase in age; only 6% of the population is over 60. As for sex distribution, we found predominance of men, who accounted for 56.07 % of the total population, while women accounted for only 43.92 %.

At present, the *Ka'aguy Poty* and *Yvy Pyta* practice a variety of subsistence strategies including a combination of horticulture, hunting, fishing and gathering—the latter of which takes place in both the "chacras" (orchards) and the "monte" (forest). Subsistence activities are combined with temporary employment in "colonias" (rural settlements devoted to the production of yerba mate, tea, tobacco and tung) and the selling of handicrafts. Some individuals receive government allowances, while others earn salaries from working as teaching assistants or sanitary agents. The money obtained from these activities allows the *Mbya* to purchase industrialized goods, such as clothing and shoes, as well as processed foods. These individuals no longer rely on the forest as their sole provider of nourishment. Changes in the natural environment, a shift towards a more sedentary lifestyle, and a processed diet high in lipids and carbohydrates have altered *Mbya reko*, leading not only to changes in culture broadly construed, but also to the culture of the human body.

Figure 1: Provincia de Misiones ©Marta Crivos, et al.

Health Assistance in Mbya Communities

In both the *Ka'aguy Poty* and *Yvy Pyta* communities, a diversity of actors play roles in the provision of healthcare. Many conditions are routinely diagnosed and cured within the domestic environment by members of the sick person's immediate social network. The extent of the participation of these actors depends on their knowledge and previous experience in the treatment of each specific illness, as well as the general availability of the necessary therapeutic resources. Many of traditional healing resources are collected in the "monte" (forest); the gathering, compounding, and application of these resources requires not only a familiarity with natural medical resources, but also a wide range of experience with disease and healing. For this reason, elder members of the community play a central role in the healing of the sick. Furthermore, there exist in every community some individuals—generally elderly men and women—who are recognized as therapeutic experts, known in the *Mbya* language as *Karai* or *Kuña Karai* (men and women respectively). The term *Karai* is used to refer to prestigious people within a community who have displayed extensive knowledge not only of natural remedies (*poã*), but also of the prayers and other rituals associated with treatment of illness. Some of these *Karai* are called *poro poãno va'e*, an expression that generically refers to someone having the ability to cure "*con remedio de yuyo*", that is, by using vegetal therapeutic resources.[11] Some of these *Karai* are also recognized as religious leaders within the community, in which case they are called *Pai* or *Opyguã.* What is distinctive about the religious leader, and confers him/her with the distinction of *Pai* or *Opyguã* is the "call" or vocation he or she receives from the gods. Unlike other specialists, the *Opyguã's* performance goes beyond the therapeutic domain; due to their extraordinary powers, they are the only ones who can lead the ceremonies related to natural, annual, and human life cycles.[12]

In addition to local healers and traditional therapeutics, the *Mbya* also have access and frequently resort to the *Aristóbulo del Valle Sanitary Unit* (US:AV:), which includes medical, odontological and biochemical staff, as well as an admission service. The *Ka'aguy Poty* and *Yvy Pyta* can also obtain medical care at the *Jardín América Hospital* (40 km away), which provides a broader range of more advanced medical services. Other alternatives include the hospitals in *Oberá* and *Posadas*, both of which are farther away from the settlements, 58 km and 147 respectively. Until 1998, there was a sanitary station in the *Ka'aguy Poty* community that provided both communities with medical attention. The staff there included a nurse, who consulted with patients presenting with routine problems, a sanitary agent (aboriginal) who was responsible for monitoring children's weight and height and distributing food supplies as provided by the Provincial Mother-child Plan. A doctor from the *Dirección de Asuntos Guaraníes* visits the communities to deliver primary health attention (PHA), but does not stay on a permanent basis. This is done with the help of a local sanitary agent who gathers the population at the local school building, consults with those who present with health conditions, performs routine clinical examinations, and measures and weighs children. The doctor on duty generally has the basic medicine (antibiotics, anti-feverals, anti-parasitics, and pain killers) necessary for the most common conditions, as well as food and supplements for those families who have under-weight children. Many members of the *Mbya* communities take the opportunity to consult with the doctor (either because of an illness or a routine check-up) when he is available.[13] It should be emphasized that, although the *Mbya* mark in their speech a contrast between traditional (*Mbya*) medicine – or "natural medicine" as some people call it – and the *jurua* medicine (that of the white people), they use both types of medical resources in a complementary, combined or alternative way when facing concrete illnesses. Nevertheless, there are some ailments that are considered by the *Mbya* as belonging to them ("propios de los Mbya"), and can only be solved by appealing to traditional specialists or therapeutic resources, as they believe them to be beyond the purview of biomedicine.

Methodology

Our ethnographic research investigated *Mbya* perceptions and categorizations of disease, and strategies to recover health. We gathered data regarding the processes of health–illness processes through the application of techniques stemming from qualitative research. We conducted semi-structured interviews aimed at recognizing the most frequent illnesses affecting individuals of different sexes and ages. We also asked individuals about their beliefs regarding the origins and symptoms of their particular conditions, as well as the diagnosis, prevention and therapeutic practices associated with their conditions. Due to the local population's concern with parasitic infections in conjunction with the results of parasitological studies and the high number of consultations regarding parasites in health centers, we decided to focus our work on gastrointestinal illnesses[14].

We closely studied cases of gastrointestinal disease by interviewing various adults who had personally experienced or participated in the events surrounding parasitic infection.[15] Further work was done with adults and children on the graphic representations of the different kinds of parasites recognized by the *Mbya* population and their locations in the human body. This allowed us to gain information about *Mbyan* perceptions of the human body, the physiological processes associated with digestion and food metabolism, and the effects of gastrointestinal parasitic illnesses upon them. As a complement to interviews, we observed daily health care practices with the purpose of identifying activities oriented towards the prevention of illness as well as recording scenes that involved the gathering, preparation and administration of traditional therapies. Members of our team also accompanied adults (both experts and laypersons) on expeditions aimed at gathering natural substances used in the treatment of parasitoses; as part of our participation in these expeditions, we gathered samples for future examination.[16]

Motivated by the local people's concern about constant re-infection and chronicity of parasitoses, ten workshops and meetings were carried out between 1999 and 2008 with the purpose of promoting discourse between the population's perspectives, knowledge and practices regarding parasites and those of the medical staff in relation to the prevention and therapeutics of enteric parasitoses. Participants were the members of the interdisciplinary team of UNLP, doctors, nurses, sanitary agents, health agents, teachers and members of the *Mbya Guarani* population and those of the *Aristóbulo del Valle* locality. The methodology applied in successive workshops combined presentations done by experts, working in smaller groups, plenary debates, and the collection of soil samples and fecal samples from domestic animals for later laboratory examination. This process allowed the members of the community to become familiar with the tools and techniques used by microbiologists as well as the procedures for the diagnosis of parasitoses by trained medical personnel. Children attending the workshops participated mainly in the graphic representation of the places and risk behavior related to parasitoses, parasite location in the human body and the images resulting from their observations in the microscope.[17]

During one of the first workshops in the *Ka'aguy Poty* community, conducted in 2000, some members of the *Yvy Pytã* community who were in attendance invited our team to a meeting where they presented their own perspectives about parasitoses. This workshop took place two days later in the community's school and was run by one of the members of the population, allowing for discussion and participation of the community as a whole as well as members of the UNLP team. The productive discussion that followed allowed us a window on *Mbya* knowledge about the body and parasites and granted us access to a set of traditional values and experiences regarding parasitoses. The result of these discussions was a refocusing of our ethnographic interviews and the incorporation of new factors into our characterization and assessment of the problems associated with *Mbyan* parasitoses.

Results

"We come with tachó": Parasites and the Digestive Process

"Acho" or *"tachó"* is the *Mbya* word for parasites in general. It is translated into Spanish as *lombrices*, *bichos* o *parásitos (worms, bugs or parasites).* According to the *Mbya*, *tachó* are normally located in the "*tripa*" (the guts), that is, inside the bowels, which are called "*gekue*" or "*tyekue*" in the vernacular language.[18] There often seems to be a distinction between "*guachu kue*" and "*gekue guachu kue*" (the large intestine) and "*gekue poi*" (the small intestine). In this respect, "*py`a*" refers to the thoracic cavity and "*tye*" to the abdominal cavity (*tye*: abdomen) and the term also involves the stomach. That is, "*py'a*" and "*tye*" are usually used to refer to the cavities and also to the organs inside them. Therefore, the names of some illnesses are also derived from these names.[19] Thus, "*tyerachy*" means "stomachache" and "*gerachy*" means "diarrhea". The suffix *-rachy* derives from "*mba'achy*", an expression used by the Mbya to designate "illness"[20].
Parasites in the digestive tract are considered to be necessary for human life and their beneficial action is closely related to the way food is transformed.

In this sense, they are seen as a basic component of the normal physiological processes of digestion. This is how some of our informants express this fact:

> "children already have tachó, we come with tachó... they are inside the person, come with you when you are born. Every child, since very little, has the 3 kinds of tachó." (CR, 65-year-old man, opyguã, YP, 2000)[21]

> "you have to have it, no matter what...it must die together,... together it has to die... (the person and the worm must die together). There you are, finished (dead) because you don't have any liquid, nothing...our parasite mother... that one is born with us...yes, it doesn't come out of the body, either, it comes with us, because we raise it since we are babies, little babies and we come with the parasite, "ñande racho", that's the way we call it, 'our parasite', the mother... "ñande racho chy" ...if you don't have any parasites, if you eat a big one (piece of food),you won't chew,... the parasite is the one who chews.... It makes molinares (it crushes it), it is like the mom of this other one (parasite)"... (MG, 30-year-old man, YP, 2000).

There would be two kinds of parasites normally living in the digestive tract: *mba'e che vera*, ("the saliva's owner"), located in the throat, and *ñande racho chy* ("our parasite mother"), located in the gut ("*tripa*"). *Ñande racho chy* produces eggs that give rise to two other forms of parasite: *tacho ovy*, green parasites, and *tacho pytã*, red parasites. The latter ones, *tacho ovy*, and *tacho pytã*, are the ones which, under certain circumstances "become agitated" and alter their behavioral patterns. When this happens, the physiological balance of the gastrointestinal system is likewise altered, giving rise to different symptoms that the Mbya refer as characteristic of parasite illness. Although parasites are located in the digestive tract, some types migrate throughout the body and may travel to the heart, stomach, through the trachea, and to the mouth. In these cases, when the "bugs" appear outside the "*tripa*", some informants express that this is another kind of illness, associated with some kind of intentional "damage" to which they give the name "*paye*". These "bugs" are the material and tangible expression of the willful damage sent by another person. Unlike other types of worms, which are treated with vegetal therapeutic resources, the treatment of *paye* worms requires the intervention of traditional specialists and the implementation of religious rituals. In the words of one of our subjects:

> "it isn't any kind of worm, there is a spiritual part, too ... it is not to be cured only with the yuyo or with... because payé is an evil spirit, there are some people who learn to do things to other people, they can directly kill another person, it is not

natural, it is sent (by another person) to do something like that ... doctors can do nothing; payé may be a worm, a spider, or any other bug that gets into your throat, into your heart … a healer must take it out … and bug eats only from your body ... sometimes it comes out of your nose, it gives you fever … and it dies, and out went the payé, when it comes out it must be killed, must be put into the fire and burnt … because if you don't, it gets into other people" (SP. 46-year-old woman KP, 2001)

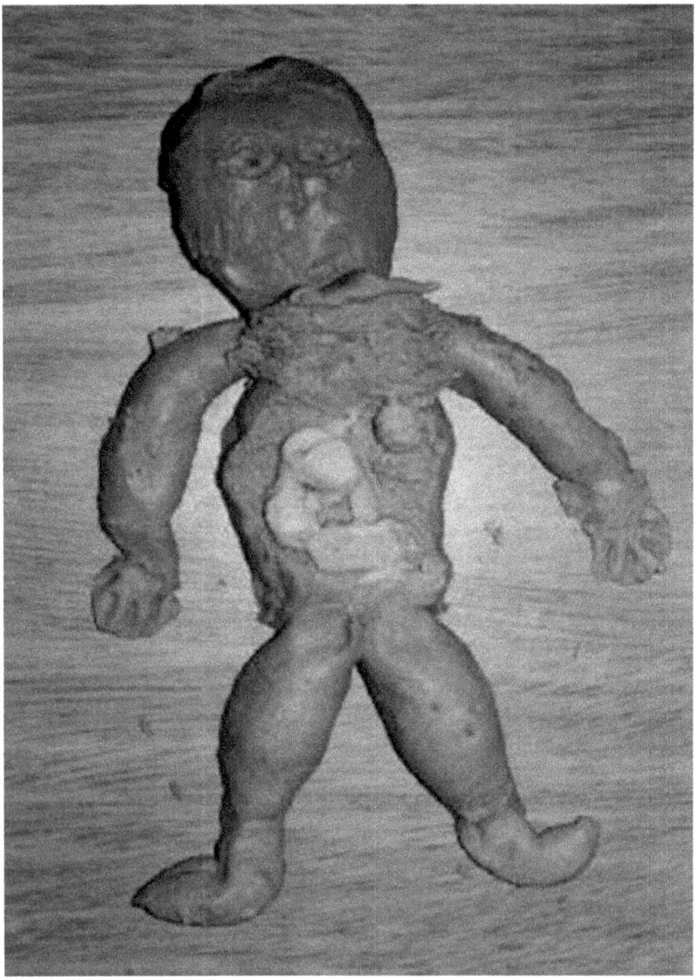

Figure 2: Bodily representation and *tacho* location in the abdominal cavity © Anahí Sy

Figure 3: Children modeling a human figure in clay © Anahí Sy

Morphological criteria are central to the identification and classification of worms found in fecal matter. Using their own taxonomic system, the *Mbya* distinguish five types of *tacho* classified according to color and size, with each type being associated with different pathogenic causes and levels of morbidity. *Tacho ovy* and *tacho pytã* are the most common pathogenic parasites found in human feces, and are either yellowish-white (*tacho morotí*) or yellowish-orange (*tacho ju*) and very small, not often reaching two centimeters. These parasites are called *"tacho i"* because of their small size. Initially, a correspondence could be established between *Enterobius vermicularis* (oxiuros) and *tacho pirirí* or *itachí*, and the two sexual forms of *Ascaris lumbricoides* as well as *tacho pytã* (male) and *tachó ovy* (female). All of these forms are macroscopic and easily observable by the Mbya. Along with these smaller worms, the *Mbya* have also identified a worm known as the *tacho guachu*, or "the solitary one," whose special features are its length and its flat shape. The *tacho guachu* is known in scientific circles as the *Taenia saginata*, and is more commonly known as the beef tapeworm. For the *Mbya*, each worm, small or large, must be treated with different therapeutic means according to its nature, size, and origin. According to our respondents:

> "... there are several kinds of parasites. There is a remedy for each of them. One, the smallest one, is cured with cangorosa[22]. The biggest one- the one that clings to the gut- is cured with ka'are[23], which is stronger. There is also another tachó. There are three all together: the smallest is called itachí, the other one -that is bigger- is called tachó sebo'i pytã, it is reddish, the itachí is white. There is another tachó, tachó morotí, of a brown color, and it is cured with parí paroba." [24] (CR, 65-year-old man, opyguã, YP, 2000)

> "....worms, right? There are longer ones and thinner ones. There are some... green ones... ovy, that is green, and there are some that are redder... red is pytã. The one that stung him more is red and it is the most serious. There are green ones, there are bigger ones, smaller ones, thinner ones, there is a little bit of everything. There are only three. Because we have three kinds, the other is kind of orange, but much, much bigger. The ones that are bad are red, the one that is green, is not. The red ones are more dangerous". (MG, 30 –year-old man, YP, 2000).

> [And the bug that falls, what is it like?] "It's like this." (She shows us the bug's size with her fingers, approximately 5-6 cm.) [And what color is it?] "Some have a real yellow color, and the other one is orange, yes, two kinds there are, see ... they are different, one is big like this, and it falls." [And that one, the bigger one, which one is it?] "It's the orange one.) [What is that called?] "Tacho, tacho ju, yellow, pyta is this one, pyta ... tacho pyta and tacho ju." [And they are together, both of them?] "Yes, two, or three, the bigger ones like this or the small ones like this." [And how many fall?] "Well, just one." (S.P, 46-year-old woman, Ka´aguy Poty, 2000)

> "Well, yes, it had more or less two parts, yes." (He cuts a branch and shows me) and a little more, like this ... yes, but when it was inside your belly it curled like this, like a ... like mboi, a snake" (RR., 27-year-old man, Yvy Pytã, 2003).

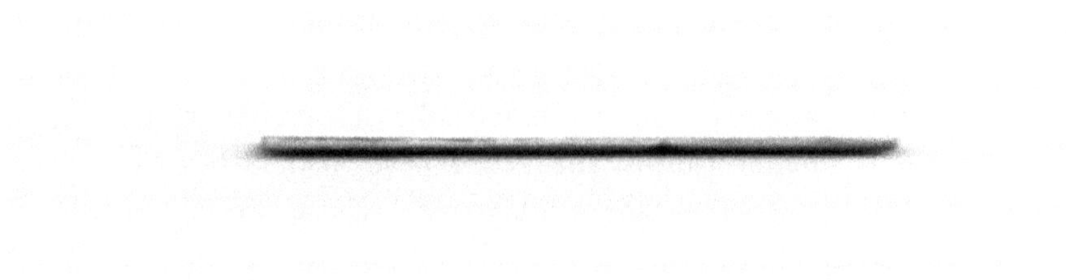

Figure 4. *tacho'i* © Anahí Sy

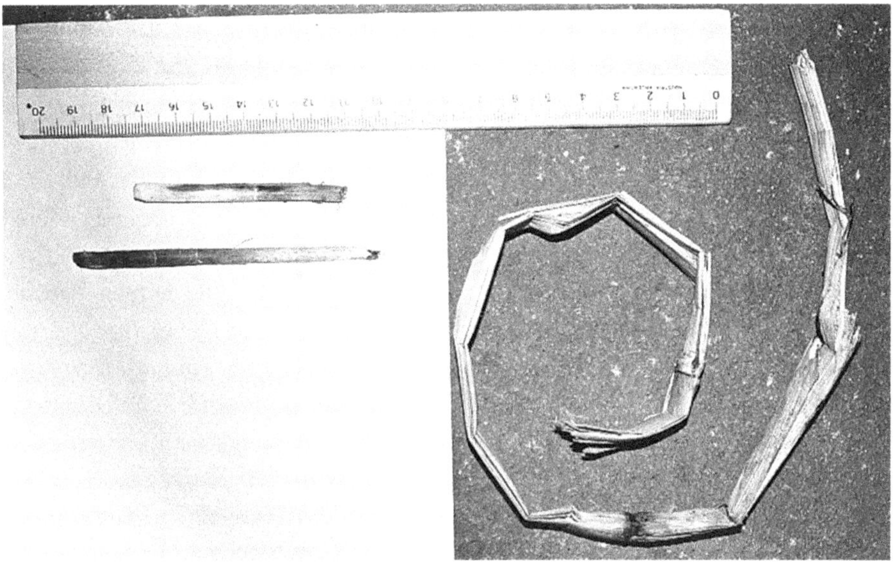

Figure 5. *Tacho guachu or "solitary"* © Anahí Sy

Being Ill: Main Symptoms of the Pathogenic Action of Parasites

Symptoms characteristic of parasitic infection are vomiting, green diarrhea (*okaa ovy*), stomach ache (*tyerasy*) pallor, loss of appetite, an increase in abdominal size, dryness of the eyeball, irritability, and disruption of sleep patterns. Diagnosis is made through observation of the individual's behavior, identification of physical symptoms, and examination of the feces for abnormal color and consistency and for the presence of macroscopic organisms. According to the *Mbya*, the symptom most frequently cited in cases of parasitic infection—that of intense abdominal pain—is produced because the parasites *"bite the guts"* and *"cling to the guts with their teeth"* or they thrash about the digestive tract. According to one respondent, *"Your belly hurts, because the parasite bites from inside* [and the child] *wants to eat candy, because the parasite likes sweet... they are heard inside the belly, it* [the parasite] *clings and it doesn't let you eat"* (E.V, 49-year-old man, Ka´aguy Poty, 1998).

From this we see that, in traditional *Mbya* culture, narratives have been created to explain the symptoms of parasitic infection and the behavior, including dietary cravings, of intestinal parasites. Children, for example, can be seen eating soil or sweet things because the parasite within it *"asks for that"*. That is, parasites are assigned human appetites and desires; they express their cravings or preferences for certain kinds of foods and their displeasure of other types of food—especially those that are considered taboo in *Mbya* culture. Because the worms *"are hungry,"* they will fight with each other *"inside the gut"* to eat up the food that the human ingests. One subject tells us that an infected child *"starts to eat things that are not normally eaten ... many times he/she eats earth too ... any kind of earth (s)he gets hold of, so mud, dust ... that's because they say that's what the body needs ... eh... vitamins or iron ... but it affects their body too ... the parasites take all that again, that is, it's the parasites that ask for that, (s)he feels like eating earth, and (s)he starts to eat earth, and drinks water, it likes sugar a lot ... fruit, corn, everything that is sweet, and meat, it must be meat, if it isn't meat (s)he doesn't eat, ... if you prepare something like corn meal (s)he won't eat ... doesn't eat anything, starts to eat earth"* (F.R, 30-year-old man, Yvy Pytã, 2003). Another man says of the worms, " *They are calm inside the gut, but if the person does not eat, it is hungry, the tachó is also hungry, the tachó shouts inside the belly, it can be heard, it bites. That is why it is not good to be hungry. The tachó likes food—it likes meat, it likes "asado"* (barbecued meat) *even more..*" (MG, 30-year-old man, YP, 2000). Moreover, when a person, whether an adult or a child, *"has too many parasites,"* the opposite happens, that is, the desire to eat is replaced by a loss of appetite. The *Mbya* say that this person *hacho rema ndo karu che* (*"He doesn't want to eat any more."*) and, therefore, he loses weight and gets weaker: the person that *"...has a lot of parasite, doesn't want to eat ... that's why he is skinny"* (E.V., 49 year-old man, Ka´aguy Poty, 1998)

Etiology and Prevention of Parasite Illness: "Tradition" and Changes in *Mbya Reko*.

As mentioned above in our discussion of the various origins of parasites according to the *Mbya*, some of worms are present in the human from the moment of birth and are necessary for human survival. Others, however, *"are bred"* when the individual transgresses certain food prescriptions. Dietary transgressions can be intentional or accidental; this is especially evident in one *Mbya* man's ideas about dietary restrictions and infants:

> "… they (infants) may eat pig meat 15 days after birth, and they say they (the infants) get dizzy, or have a fit, because of parasites, that's why it is forbidden ... the father can't (eat pork) either;, that is, the baby can be attacked if the father eats and sometimes, well, gets diarrhea but there are things that only the opyguã know what it is, that only they can save the children…"
>
> "... my dad says it comes from the mother's way of eating ... when my dad asked what I had eaten excessively, we became aware it was pig's meat ... when she (my mother) was pregnant we used to eat the meat a lot ... and no, that meat mustn't be eaten... the baby already had it, was born with it ... when you are pregnant, any food you can't eat ... it makes it (the parasite) breed (inside the body of the fetus) ..."
>
> "Yes, it breathes, like this, inside, then the parasite is produced... If you cook coati, you hunt the coatí, and this little baby doesn't eat yet, then it comes in with the wind, and it gets in, and there the parasite produces... (it breeds because of) the smell of the foods, that's the way it gets into the little ones, it comes to them, with the wind, and he breathes, the coatí's smell, you make the barbecue and it comes with the wind, it's a smoke, let's say, then you swallow it and it is produced..." (MGo. 30.year-old man, Yvy Pytã, 1999).

As evident from the statements above, the parasitoses in toddlers under the age of one can be attributed to three main causes: the consumption of the meat of "forest animals" by women during pregnancy; the exposure of children (who haven't been "baptized"[25] yet and as such are uninitiated) to the smell given off by the cooking of "woodland animals" or the mixing food that should be eaten separately, both of which constitute "bad eating." The probability for a little child to get parasites is closely related to taboo transgression during pregnancy and the first months of the child's life, when the baby is considered to be "new" (*mita pyta*). During gestation, both parents must closely adhere to a set of restrictions centered on the food they consume. In this regard, forest animals, especially the *coatí* and *kure* (wild pig) are considered to be the most dangerous of foods for the infant. One informant told us about two of his children who have had parasites since they were born; according to him, they are infected because his wife ate *kure* while she was pregnant and, although she was not affected, this type of *"meat makes the worm breed, ...makes it grow inside the little one".*

Although the children were currently being breastfed, "*s)he* (one infant) *was skinny, because (s)he also vomited and threw up the milk, (s)he was pale"* (F.R, 30-year-old man, Yvy Pytã, 2003).

Just as infants are not to consume the flesh of wild animals before they are baptized, so too are they not to inhale the aroma of wild meat as it is being grilled. Thus the second cause of parasitic infection in toddlers, like the first cause, is related to the transgression of dietary and cultural taboos. Although the wind is described as the agent enabling the "activation" and growth of parasites that exist "naturally" within the infant's body from its formation in utero, it is nevertheless the parents' responsibility to protect the child from exposure to the smell of grilling meat and the possible parasitic infection that may result. The role of the parent in protecting the child is also evident in the third cause of infection: inappropriate food choices and regimen. Feeding a child forbidden food that is too *"heavy"*[26]for his/her age, as well as stopping breastfeeding and simultaneously giving different types and flavors of foods potentially "activates" latent parasites within the human body and triggers symptoms of parasitic illness. The *Mbya* generally argue that *"tachó is easily solved;"* however, if a child has a great number of parasites or if the parasitic infection is allowed to proliferate over a long period of time, the child's condition may worsen. In these cases, children become *"weak", "not lively," "don't want to grow,"* and *"they get fits, dizziness; they may faint, they fall."* That is, these children become vulnerable to other illnesses and suffer from weight loss and delay in their growth.

To sum up, there exist certain food taboos, established by *"the ancients"* (the *Mbya* ancestors), that regulate the consumption of some kinds of food considered to be harmful; the disregard of these taboos produces parasites that can damage health and cause death. The pathogenic action of the *tachó* is stimulated not only by the breaking of taboos, however, but also by the consumption of industrially processed sweets:

> [You had also told me that your daughter had tachó?] "Yes, tachó, parasite, that was last week, she had, parasites." [And how did you notice that she was ill?] "Because she cried, she almost didn't stop … we immediately knew it was that. [And why does that get to children?] "Well, I myself don't know how this parasite comes, how it does to come, because I sometimes bring a little sweet cake and we give her, that's how it comes". [And the parasite is already in the body, or does it come in?] "No, it's inside. When we are born, when one is born, it's already there…" (B.D, 28-year-old man, Ka´aguy Poty, 2003)

Beyond the breaking of dietary taboos established through *Mbya reko* and the consumption of processed sweets, different types of parasites may also enter into the human body from the environment, particularly the water and the soil. The *Mbya* are aware of these problems as well.

> [Do children and adults have it as well?] "Yes, they do, everybody ... when they start to walk they already have ... and children too ... they stand up, like this, they walk and drink water." [And before they walk, they don't drink water?] "No, water is not drunk yet, we don't give them, they have milk, they drink milk, ... milk; when they drink milk they don't have tachó... because they have water, with the water only it comes, sometimes they take it raw, like this, or first ... that kind, runoff water, as it is from a stream it shouldn't be drunk, but we sometimes drink it, just like that, runoff, from the stream, ... an there they are, already." (S.P, 46-year-old woman, Ka´aguy Poty, 2000)

> "... there, in the (water) well is where they get everything, the children get ill, that well is too dirty ... " (A.R, 28-year-old woman, Yvy Pytã, 2003).

The problem of parasitic infection through water arises mainly in children who are weaned from breast milk; they start to walk and drink water (from the stream or runoff). The *Mbya* are advised to boil water before it is used, and are warned not to drink it "raw". Parasites that enter the body through drinking water are considered to be pathogenic, unlike those which "are born" with the individual.

From the results of our research, the main approaches to the prevention of parasitoses in the *Mbya* communities include:

1. Keeping infants on breast milk for as long as possible
2. Observing Mbya dietary restrictions that date back to the times of the "ancients."
3. Avoiding "heavy" foods, primarily of industrial origin, like sweets.
4. Boiling runoff or stream water before use or consumption.

The theory that parasite infection can be caused by contaminated water reflects the influence of modern biomedicine on traditional *Mbya* culture. Accounts provided by the *Mbya* show a correlation between this shift in *Mbyan* perceptions of the causes of parasites and the information coming from health programs and from the activities of sanitary agents, who are trained in the scientific method, and who assert that the causes of intestinal parasitoses are contaminated soil, water, and food—all of which are external to the human body and odds with the traditional *tachó* of *Mbyan* culture.[27] In this sense, sanitary agents, who are a nexus between traditional and biomedical cultures, have a central role as intermediaries in the population's gradual access to scientific knowledge and practices. As an example, let us examine the following statement:

> "...because I think that's true, ... because there is my little daughter, who has parasites, and my wife said that is because I ate coatí's and pig's meat too soon, and that's really true, because it has been proved, you see. For me it is, and I know something about cleanliness too." (AD, 34-year-old man, Aboriginal Sanitary Agent, 1999)

This informant acknowledges the need for parents to respect food related taboos that have to be observed while the child is "new" (*mitã pyta*), and reinforces their validity ("that has been proved") while at the same time incorporates another factor—lack of hygiene—which is prioritized by the biomedical sector as the main cause of enteroparasitosis.

Final Considerations

We have so far developed the most outstanding aspects of *Mbya* practices and representations about parasites and parasitoses, mainly stressing native considerations about the causes of the pathogenic activity of "worms," mainly food taboo transgression. We also stated that this kind of explanation is strongly anchored in ancient, ancestor-transmitted "tradition," which focuses on harmonic relationships between human beings and other organisms in their environment.[28] Our research has also revealed the recent incorporation of modern biomedical explanations for parasitoses and the health risks they entail into traditional *Mbya* culture, a phenomenon that can be attributed to ever-increasing interactions between public health agents and physicians—as well as the modern health systems they represent—and the *Mbya* people themselves.

Analysis of Mbya conceptions about the relationship of taboo infringement and sickness highlights the value ascribed by *Mbya reko* to the harmonic relationships between human beings and the environment—as well as the continuity of ancestral practices as a guarantee of balance. The infringement of the various taboos that regulate the interactions between entities, such as worms and their hosts, appear as one of the main causes for the illness. Representations of parasitic illness and its consequences reflect an ecological approach to viewing human health—an approach that depends on maintaining a balance between different types of organisms. In this sense, the conceptualization of illness as a "time dis-adjustment to the environment" proposed by Jacques May[29] corresponds to the *Mbya* perspective about the appearance of the parasitical illness. From this relational perspective, the appearance of illness serves as a signal to an imbalance in the environment that is reflected in the body—one that can alert researchers and public health officials to what has to be done to repair the imbalance.[30] As Young (1976) points out, medical beliefs and practices have not only a practical or instrumental value but also a symbolic one that can communicate and reinforce concepts about the human body and the greater macrocosm of which it is a part.[31] *Mbya* representations of parasitoses, therefore, provide us a path into traditional conceptions about the individual person, his or her social relationships, and the various interactions between the human and non-human entities in the primary *Mbya* environment: the Paranense Forest.

Dis-adjustments between the *Mbya* and their environment can be attributed to sedentary modalities in intensely exploited spaces, a decrease in traditional horticultural practices, the scarcity of wild animal resources, and the replacement of traditional food sources with those of industrial origin. All of these are contributing factors to the increased vulnerability of the *Mbya* population to intestinal parasitoses. *Mbyan* perceptions of biological and ecological changes agree with the findings of scientific research. Modern scientific studies have shown that in numerous South American indigenous communities, changes in the ecosystem, in traditional subsistence strategies and in dietary regimen are the main factors for the virulence and chronicity of intestinal parasitoses—particularly when combined with nutritional deficiency, an endemic problem in many of these traditional cultural groups. Numerous studies that focus on the co-evolution of parasites and their hosts point out that South American aboriginal populations show unique physical health and illness patterns, suggesting a high degree of human adaptation to parasitoses.[32] This research indicates the presence of high antibody levels in these populations, due to their exposure to numerous infectious agents, including intestinal parasitoses, protozoa, and bacteria.[33] Nevertheless, gradual changes in the traditional indigenous culture can have a negative influence on parasite-host relationships, increasingly making the presence of parasites dangerous to the body.[34] On the basis of these results, our researchers are now focusing on the impact of cultural change among indigenous populations with the intent of evaluating the relationships between rates and type of parasitic infection and shifts in settlement patterns, changes in diet, stress levels, and increased contact with modern cultures. We are also considering the influence of modern biomedicine, the extensive use of antihelmintics, as well as environmental modifications which supposedly influence parasite-host relationships, causing a higher susceptibility to infection.[35]

The *Mbya* perspective on parasitoses has some points in contact with the perspective proposed by population ecology, which argues for a complex and multi-valent consideration of the conditions that make parasitoses possible. The *Mbya* ultimately rely on their belief systems to interpret and manage intestinal parasitoses. These views contrast sharply with those from classical epidemiological studies, which focus on broad identification of the issues and an "*a priori*" definition of over-generalized "risk factors" without considering the historical depth of the relationships existing between indigenous populations and their complex ecological environments or the unique experiences and factors relevant to each individual's daily life.[36] Such broad-based approaches as those espoused by classical epidemiology do not allow for hypotheses that consider the interaction of cultural and environmental factors in illness distribution in different groups within the same population.[37] We argue that the relationships between ecological, genetic, cultural, sanitary and economic factors that affect a particular populations' health should be considered in greater depth, an approach that is by its very nature interdisciplinary and systemic.

Finally, some considerations about our interdisciplinary work experience with the *Mbya* population. Working with parasitologists and biological anthropologists allowed us to identify both shared and exclusive aspects in scientific *vs.* local explanations about the identification and categorization of parasites and the environmental conditions which make parasite development possible. The way parasitoses are perceived was explored in workshops where members of the indigenous population, scientists and sanitary staff participated. Discussions about varying perspectives allowed for the identification of different recognition thresholds of the impact of parasitoses on human health. Moreover, the contributions of ethnobotanists regarding alternative treatments and prevention strategies led to a positive appraisal of traditional knowledge and handling of natural resources. Intercultural exchanges favored by the ethnographic approach opened a pathway for the positive exchange of ideas between biomedicine and traditional groups. From the first perspective, parasitologists and health staff acknowledge the *Mbya* community's limited knowledge of parasites and their environmental causes—an information gap that has allowed environmental "risk factors" to proliferate. As a prevention strategy, modern scientists argue that the environment must be regulated and that the *Mbya* must change many of their traditional habits in order to avoid parasitic transmission and infection. The *Mbya* people, on the other hand, claim that parasitic infection has nothing to do with these issues, but instead with the breaking of dietary taboos. In this sense, they emphasize the need for respecting traditional guidelines which, ultimately, will prevent parasites from becoming pathogenic. Consequently, both sectors diverge on their opinions about prevention. Native sanitary agents, as members of both the biomedical and traditional *Mbya* worlds, integrate both explanations into their view of parasitic infection and prevention.

Most workshop participants, including local people and sanitary-medical staff, at first believed that parasites were not an issue of great concern, that worms were not only tolerated, but a normal part of the human life-cycle. Through the workshops hosted by our group, the *Mbya* population was educated about the biological aspects of parasitic infection and reproduction and their consequences for the human host. From this point forward, while we were performing fieldwork we began to notice a greater concern on the part of the *Mbya* population about lack of hygiene and parasitic illness. Local sanitary medical staff, in turn, became more receptive to *Mbyan* perspectives on parasites, allowing them to find medical approaches to parasitic control that would mesh more closely with *Mbya reko*. Taking into account the impact of parasitic pathologies on the overall health of the region, it seems relevant to us that the workshops, through dialogue, allowed for discourse between two divergent communities and widened the each sector's knowledge about parasites and associated illnesses.

Acknowledgments

This research was supported by National Counsel of Scientific and Technical Research (CONICET) and University of La Plata. We are indebted with all members of Mbya communities of Ka´aguy Poty e Yvy Pytã for their warm hospitality during our fieldworks. Also, we are especially grateful for their engagement in the development of projects oriented to know and solve health problems affecting the local population, in which we work together.

Notes

1. See E.E. Evans-Pritchard, *Witchcraft, Oracles, and Magic among the Azande* (Oxford: Clarendon Press, 1937). Nurit Bird-David, "Animism Revisited: Personhood, Environment, and Relational Epistemology," *Current Anthropology*,vol. 40, (1999), 67–91.
2. Navone, G.T, M.I. Gamboa, E. Oyhenart, A.Orden, "Parasitosis intestinales en poblaciones Mbyá-Guaraní de la Provincia de Misiones, Argentina: aspectos epidemiológicos y nutricionales," *Cadernos de Saúde Pública* 22 (2006), 5.
3. Continuing Education Projects: "Strategies for the Integration of the Mbyá-Guaraní Community at Ka'aguy Poty, Valle del Cuñapirú, province of Misiones, in Practices for the Diagnosis and Prevention of Parasitoses" 1999–2000. "Parasitosis and Parasite Illness in Suburban and Rural Villages in the Area of the Reserva Privada UNLP Valle del Arroyo Cuña Pirú, province of Misiones: Strategies for their Diagnosis, Treatment and Prevention" 2002–2003. "Intersectorial Building of Strategies for the Handling and Control of Parasitoses. A Trial at School N° 172, in Aristóbulo del Valle, Misiones." 2006–2007. Directed by Dr. Marta Crivos and Dr. Graciela Navone. Coordination: María Rosa Martínez. Aproved by Resolution of the Superior Counsel of the Universidad Nacional de La Plata.
4. V. Assis & I.J. Garlet, "Análise sobre as populações guarani contemporâneas: demografía, espacialidade e questõs fundiárias," *Revista de Indias* LXIV: 230 (2004), 35–54.
5. ENCUESTA COMPLEMENTARIA DE PUEBLOS INDÍGENAS (ECPI). 2004–2005. Resultados provisionales. Provincia de Misiones Ministerio de Economía y Producción. Secretaría de Política Económica. Instituto Nacional de Estadística y Censos. Available in: <www.indec.gov.ar>.
6. Ivori Garlet, Ivori, *Mobilidade Mbya: historia e significacao* [Tesis de Maestría]. Pontificia Universidade Católica do Rio Grande do Sul. MS; Bartolomé, 1978, Gorosito Kramer, 1982;

7. C. Remorini, "*Caminar a través del monte. Una aproximación a la movilidad Mbya en el pasado y en el presente*," in *Publicación Especial de las II Jornadas sobre Poblamiento, Colonización e Inmigración en Misiones*, 309-326.
8. Crivos, Marta, M. Rosa Martínez, Graciela Navone, M. Lelia Pochettino, Patricia. M. Arenas, Celina Digiani, Laura Teves, Carolina Remorini, Anahí Sy, Carolina Illkow y Nicolás. Delorenzi. "Ethnobiology of the parasitoses: the case of two Mbyá-Guaraní communities (Province of Misiones, Argentina)," in *Ethnobiology and Biocultural Diversity*, eds. John R. Stepp, Felice Swyndham and Rebecca K. Zarger, (Athens, GA: University of Georgia Press, 2002), 258–269.
9. Carolina Remorini & Anahí Sy, "El valor del monte en el proceso de endoculturación Mbya. Una aproximación etnográfica," *Actas del XXIII Congreso de Geohistoria Regional*, Oberá, Misiones: Facultad de Artes. Universidad Nacional de Misiones (2003), 115–127.
10. María L. Pochettino, María R. Martinez, Patricia M. Arenas, Marta Crivos, Graciela Navone, Celina Digiani, Laura Teves, Carolina Remorini, Anahí Sy , Carolina Illkow y Nicolás Delorenzi. "Fitoterapia Mbya: análisis y evaluación del tratamiento de las parasitosis," *DELPINOA, Journal of Ethnobotany* 46 (2003), 301–310.
11. M. R. Martínez, M. Crivos & C. Remorini, "Etnografía de la vejez en comunidades Mbyá-Guaraní, provincia de Misiones, Argentina," in *Il Vecchio allo Specchio. Vivere e curare la vecchiaia nel mondo, Vol. 4*, eds. Guerci y Consigliere, (Geneve: Erga Edizione, 2002).
12. C. Remorini, "Las relaciones intergeneracionales en la vida cotidiana. Sobre el rol de los abuelos en las actividades de cuidado infantil en comunidades Mbya (Misiones, Argentina)," *Actas del VIII Congreso de Antropología Social, Simposio 14* (2006): Cultura y envejecimiento. Abordajes multi e interdisciplinarios. Universidad Nacional de Salta, CD-ROM. Salta: EDUNSa.
13. A. Sy & C. Remorini "Hacia un abordaje integral e intercultural de la salud de los niños Mbya. Contribuciones de la investigación etnográfica y desafíos para la gestión pública," in Noceti M.B (Comp.) *Oportunidades. Caminos hacia la protección integral de los Derechos del niño*, (Bahia Blanca: Libros en Colectivo, 2009).
14. We included in this category those illnesses that were recognized and diagnosed by the local population whose symptoms correspond to the pathologies classified by biomedicine under this name.
15. Anahí Sy, *Estrategias frente a la enfermedad en dos comunidades Mbya Guaraní (Ka'aguy Poty e Yvy Pytã, Provincia de Misiones). Aporte del estudio de casos a la investigación Etnográfica de los procesos de Salud-enfermedad.* (Tesis de Doctorado). Facultad de Ciencias Naturales y Museo. Universidad Nacional de La Plata, 25 de abril de 2008.
16. Crivos et al., "Ethnobiology of the parasitoses," 2002; Crivos, Marta, M. Rosa Martínez, Graciela Navone, M. Lelia Pochettino, Patricia. M. Arenas, Carolina Remorini, Anahí Sy, Laura Teves, M.Inés Gamboa y Lorena Zonta, "Puesta en común de saberes y prácticas sobre las enteroparasitosis (Valle del Cuña Piru, Misiones)," Equipo de

Etnobiología, Museo de Antropología, Facultad de Humanidades, Universidad de Córdoba. Edición en CD (2006); Pochettino et al, "Fitoterapia Mbya," 2003.

17. Crivos et al., 2005.
18. León Cadogan, 1992 Diccionario Mbyá-Guaraní-Castellano. Biblioteca Paraguaya de Antropología. Vol. XVII. Fundación "León Cadogan", Asunción, 183.
19. *Ibid.*
20. This expression´s meaning comes from mba'e (thing, entity; what causes pain) and achy (imperfect, pain) (Cadogan, 1992: 103).
21. Informants' names are only given by their initials to preserve their anonymity.
22. Maytenus ilicifolia Mart. ex Reissek (Celastraceae)
23. Chenopodium ambrosioides L. (Chenopodiaceae)
24. Piper mikanianum (Kunth,.) Steud (Piperaceae)
25. By baptism we mean the ceremony of giving the child a sacred name (Ñemongarai). It is made once the child acquires two skills: talking and walking straight. For the Mbya, the soul is identified by the name, and the ability to speak constitutes the subject's vital principle and it will accompany him to his/her death. The importance of this ceremony in relation with the vital cycle is shown in the fact that onlñy after it is the child recognized as a full member of the teko'a (the community). See Remorini *Aporte a la caracterización etnográfica de los procesos de salud-enfermedad en las primeras etapas del ciclo vital, en comunidades mbya-guaraní de Misiones, República Argentina.* -1a ed. – (Tesis de Doctorado). 11 de abril de 2008. La Plata: Universidad Nacional de La Plata. Available in: www.unlp.edu.ar/editorial..
26. Based on their effect on the body, food is classified as "heavy" or "light". This is not a fixed classification, since there are some kinds of food that are considered to be "heavy" only if they are consumed when the body is vulnerable, for example during the transition between diffeent stages of the life cycle (baptism, first menstruation, puerperium) or when the individual is ill. For example, while children over one year of age who have been baptized are allowed to eat meat, mix some kinds of food and/or eat "heavy" food on some occasions, the Mbya perception of vulnerability of children who have not gone through this ritual justifies their stressing on unweaned babies protection against the multiple risks related with feeding. "Light" food is generally vegetable in origin, while "heavy" food is that of animal origin and /or coming from the "jurua" (white people). Among the later are those kinds of food belonging to the Mbya, when they are consumed in wrong instances, and also industrialized food (Remorini, 2009).
27. Crivos, et. al., "Ethnobiology of the parasitoses," 2002; Sy, *Estrategias* ,2008.
28. Mbya reko is characterized by the harmonic relationship between the natural and supernatural orders. This is based in the observance of the principles stated by Ñamandu Ru Ete (Our Father, the Sun) at the beginning of times. In contemporary Mbya's speech there is always a reference to ideal rules of life, established in times of "the ancients", which is idealized in their accounts as those times when there were not so many illnesses, when one could "live peacefully" as there were better lands for cultivating, which made reciprocity possible as production could be shared. Imbalance is expressed in their

present accounts by making reference to several factors like having lost the forest, the scarcity of land, social tension and conflicts, worsening and/or appearance of new illnesses, among others of which they blame on the contact with the white people. Therefore, the changes in the way of living of "the new ones" (younger generations) as opposed to "the ancient ones" (elderly people, their ancestors) would account for the higher prevalence of illnesses in these communities.

29. Jacques M. May, "Ecología de las enfermedades humanas," in *Estudios sobre Ecología Humana. Estudios Monográficos III*, (Washington, Unión Panamericana, 1958), 91–112.
30. Sy, *Estrategias*, 2008.
31. Allan Young, "Some Implications of Medical Beliefs and Practices for Social Anthropology, "*American Anthropologist*, 78 (1976), 5–24.
32. Salzano, Francisco M. y S. M. Callegari-Jacques, *South America Indians. A case study in evolution. Research Monographs on Human Population Biology*. (Oxford: Clarendon Press, 1988).
33. *Ibid.*
34. *Ibid.*; Confalonieri, et al. 1991
35. Sy, *Estrategias*, 2008.
36. Smith, H.M., R.G. de Kaminsky, S Niwas, R.J. Soto y P.E. Jolly. "Prevalence and Intensity of infections of Ascaris lumbricoides and Trichuris Trichiura and associated Socio-demographic Variables in four Rural Honduran Communities," *Mem Inst. Oswaldo Cruz*, 96:3 (2001), 303–314; A. Fontbonne, Freese-de-Carvalho E., Acioli M.D., Sa G.A y Cesse, E.A.P., "Factores de risco para poliparasitismo intestinal em uma comunidade indígena de Pernambuco, Brasil," *Cad. Salude Pública*, 17: 2, (2001), 367–373; N. Labiano-Abello, "Epidemiology of Hookworm Infection in Itagua, Paraguay: a cross sectional study," *Mem. Inst. Oswaldo Cruz*, 94:5, (1999), 583–586. R. A. Miranda., et al., "Prevalencia de Parasitismo intestinal nas aldeias indígenas da tribo Tembé, Amazonia Oriental Brasileira," *Rev da sociedade Brasileira de Medicina Tropical* 32:4 (1998), 389–393; R. A. Miranda, Xavier, F.B y Menezes R.C., "Parasitismo intestinal em uma aldeia indígena Parakanã , sudeste do estado de Pará, Brasil," *Cad. Salude Pública*, 14:3 (1999), 507–511; Anderson, et al., 1993.
37. Remorini, *Aporte*, 2009.

PARASITES AND THE POSTCOLONIAL: WILLIAMS SASSINE'S *SAINT MONSIEUR BALY*

Charlotte Baker

Francophone writer Williams Sassine was born in Kankan, Guinea in 1944. His childhood was influenced by his Lebanese Maronite Christian father, his Guinean Muslim mother, and his studies at the colonial primary school in Kankan and at the local Koranic school.[1] However, it was the stutter that afflicted him, rather than these influences during his childhood, which led him to a life of relative solitude.[2] In 1961, during his final year at the lycée in Conakry, he participated in the first major student strike against the already repressive policies of then President of Guinea, Sekou Touré. The confrontation led to the arrest of many of his classmates and their imprisonment at Camp Alpha Yaya, providing the initial impetus for Sassine's voluntary exile.[3] He studied in France for five years before leaving to work as a Mathematics teacher and Headmaster in a number of West African countries whilst pursuing his writing. Fleeing inter-ethnic conflict in Mauritania in 1989, Sassine returned to a precarious life in Guinea, where he held the post of Chief Editor for the fortnightly news magazine *La Guinée-Djama*, contributed to the pedagogical review *l'Educateur*, and wrote for the satirical paper *Le Lynx*, commenting on the country's slow recovery from Sékou Touré's dictatorship. In 1983 Sassine was elected Chevalier des Arts et des Lettres en France and then Officier des Arts et des Lettres en France in 1993. He died in Conakry on 9 February 1996.

Sassine published four novels during his lifetime; *Saint Monsieur Baly* (1973), *Wirriyamu* (1976), *Le Jeune Homme de sable* (1979) and *Le Zéhéros n'est pas n'importe qui* (1985). He also produced a number of plays including *Légende d'une vérité* and *Tu Laura* (1995), the unfinished *Les indépendan-tristes* (1999), as well as the collections of short stories *L'Alphabète* (1982) and *L'Afrique en morceaux* (1994). A fifth novel, *Mémoire d'une peau* was published posthumously in 1998. Sassine's work has been described by Tierno Monénembo as 'the most sincere and original of his generation'.[4] A common trait of Sassine's novels is a focus on the underside of a society in crisis and attention to the human body and its degeneration, which functions as a metaphor for societal decay. Although Sassine shares the concerns of other francophone African writers such as

Ousmane Sembène with the serious problems that confront the African continent, he rarely engages in social or political polemics, focusing instead on the often-solitary individual struggling to combat the actuality of his or her condition in colonial or postcolonial Africa.[5] For Sassine, these evils originated in Islamic conquest, followed by the European colonisation that culminated in the political, economic and cultural alienation of Africans.[6]

Each of Sassine's novels centres on a quest for the recovery of a lost or concealed identity, involving a descent into a hellish situation, during the course of which the protagonist transgresses prevailing rules in order to demand a new order of things. Characteristic of this is the struggle of the eponymous protagonist of *Saint Monsieur Baly* to set up a school for the children of the poor and disenfranchised, and his attempts to deal with the everyday realities of life in postcolonial Guinea. Monsieur Baly is informed of his enforced retirement in a letter, which he reads disbelievingly again and again:

> Monsieur Baly, you are worn out, you are no longer of great use to us; your role as educator is too weighty for you at present; your teaching methods have always been archaic; we reproach you especially for not having been able to work with your new European colleague [who] embodies cooperation with France, and cooperation is the new face of fraternity in the Francophone world. What is more, you advocate the pure and simple suppression of teaching in French in favour of national languages. When your European colleague presented you with his reforms for a rapid dissemination of his language, our shared language of culture, you uttered this unfortunate phrase: 'Africa is not a testing ground, even for your pedagogical rantings: the Sahara should do you'. That, Monsieur Baly, is politics. That's serious.[7]

Baly is criticised not for incompetence, but for initiative, as his approach to teaching collides with the colonial vision of education, which is presented as a fraternal gift to the colonised from a more 'civilised' country. Sassine's choice of the school as a focus for the novel is by no means accidental, given the setting of novels including Bernard Dadié's *Climbié* (1956), Cheikh Hamidou Kane's *L'aventure ambiguë* (1961) and Camara Laye's *L'enfant noir* (1954) in this same fictional space, and Sassine's own experiences as a teacher and Head master.[8] As Aedín Ní Loingsigh remarks:

> ...quite apart from its significance as a figurative means of reflecting upon the more brutal excesses of the colonizing project itself, the colonial school is frequently an important stage for the initial encounter of the colonized student with the

> foreignness of the colonizing culture; a traumatic and alienating experience from which they will never recover.[9]

As in the novels mentioned above, the microcosm of the school in *Saint Monsieur Baly* allows Sassine to examine closely the contradictions that have shaped and continue to afflict much of the African continent. The narrative follows the charismatic Baly who, with the assistance of François the leper and the blind Mohamed, overcomes the numerous obstacles in his way to founding a new school for outcasts from society. He proposes to shake up the postcolonial education system, rejecting its principles of selectivity and elitism to open up education to all, having realised that 'there were millions of little children and illiterate adults that he could serve.'[10] He describes his task vis à vis the students of tomorrow as a process of 'tearing their souls from the shadows of ignorance, from the long series of lies and humiliations inherited from years of colonisation.'[11]

Baly draws up the plans for the school, oversees the building work and designs the curriculum, but although he realises he has three major obstacles in his way: 'the white administrator, the weather and poverty,'[12] he also faces unexpected challenges. Some parents oppose the involvement of their children in the building of the school when they could otherwise be working in the fields. He faces the disdain of shopkeepers and businessmen who fear a rise in the number of young people able to read and write, who will then challenge them on the arbitrary prices they put on their goods. Even the accountant in whom Baly entrusts the financial management of the project takes off with the school coffers. Yet he overcomes these and other challenges. When promises of equipment from the town hall fall through, Baly undertakes the groundwork himself. The education authorities fail to provide funds for the project and so Baly pours his life savings and pension in the project and leads discussions with the parents of the school children to agree the level of fees. When money is short, he finds alternatives, describing a prize giving ceremony in which his old books and textbooks are presented as awards. His vocation and belief sustain him in the face of these challenges, as well as the children in whose education he takes such pleasure: 'The *joie de vivre* that I read on the faces of the young pupils during recreation is my only sustenance.'[13]

What is remarkable about Sassine's *Saint Monsieur Baly*, however, is not the carefully crafted portrayal of the protagonist Baly, nor the political significance of the novel's focus on schooling, but the infestation of the narrative with parasites, including ants, fleas, flies, cockroaches and mosquitoes. The variety and force of parasitic imagery in the novel is intimately related to Sassine's critical perspective on postcolonial Africa, for the insects that pervade *Saint Monsieur Baly* emphasize the challenges posed to Baly's project by the physical environment in which he attempts to work, render explicit the marginalisation of the central characters in the novel, highlight the characters' various psychological states and interfere with the narrative form of the traditional Western novel.

Ousmane Kaba highlights the centrality of animalistic metaphors to Guinean fiction, contending that the diversity of animals and insects in these post-independence novels 'constitutes a metaphor for the largest social class of the country: that of the poor and disinherited'.[14] Indeed, Sassine focuses on the underside of society in his fictional writing and the majority of his characters are outsiders or rebels, beggars, madmen, lepers, albinos and the blind. Among them, the deaf and elderly Bandia of *Le Jeune Homme de sable*, the old beggar Mohamed in *Saint Monsieur Baly*, Sadou, the deaf-mute gatekeeper in *Mémoire d'une peau*, and the terminally ill Kabalango in *Wirriyamu*. The insects that pervade *Saint Monsieur Baly*—flies in particular—symbolise Baly's fall into the world of the marginalised and excluded, even if it is he who takes the first step towards that world in his encounter with François the leper:

> The image of the fat blue fly with its bulging abdomen filled with blood clouded his vision for a moment. He quickly pulled himself together and decided to approach the purulent man; he no longer pitied him. He was drawn towards this walking putrefaction by a psychological identification and an obscure wish to put into action the lessons of charity he had long taught.[15]

Although Baly finds a place in this underworld, the other inhabitants of the town consider his social death complete when he becomes involved with the poor and the sick. As Kaba elucidates, 'Monsieur Baly and his cortege of the sick, beggars, flies and microbes represent impurity, the tainting of the town.[...] In the collective conscience of the inhabitants, the fall of the old teacher into the world of beggars distances him from mankind and from God, and locates him in the world of the marginal'.[16]

The accumulation of flies around Baly reflects his moral and physical decline: moral because the former Headmaster has lost his social status as a consequence of his forced retirement and replacement with a younger man;

physical because Baly finds himself infected with the leprosy that will eventually lead to his death:

> 'Monsieur Baly, believe me or not, you have leprosy'. Then he held his test tube up to the light and pointed with his finger to the old teacher's blood: 'This blood looks quite normal, but it is packed with viruses; I saw them under the microscope.[17]

The microbes which cause his leprosy are the smallest of living organisms described in the narrative, but they are among the most feared. As Kaba remarks, a proliferation of references to 'microbes', 'germs' and 'viruses' are found in the novel, on each occasion linked with leprosy.[18] In descriptions of 'the microbes which swarmed about in his blood,'[19] the movement of the microorganisms recalls the incessant buzzing of the flies, and they are just as vicious, irritating and voracious. The aggressive nature of the flies that appear throughout the novel is made clear in the description of the bluebottle which pesters Baly: 'The big bluebottle turned on the attack, and swooped down on Monsieur Baly, taking particular pleasure in walking on his freshly shaven head.'[20] The flies harass Baly, '[making] the most of his weakness to attack him and soon clustered around his large bald head, his ears and any part of his body left uncovered. The more he squashed, the more that came.'[21]

Flies in the novel inhabit the bodies of the most marginalised members of society, among them the beggar François, who terms himself 'the rotten-mad-leprous-damned man.'[22] In the opening pages of the novel, the reader is party to a suicide attempt by François, who finds no reason to prolong his solitary and unhappy existence:

> So You, my God, You in the name of whom I lived as a good man, You abandoned me and You took his side: my wife and my child died shortly after suffering simple headaches; You gave me leprosy, look at my hands and my feet and my face; I am so rotten that I do not even notice my own smell; the creditors seized all my goods and made me homeless. My fellow citizens, young and old alike, flee, crossing themselves at the sight of me; those who dare approach me spit on me or throw stones.[23]

François is tormented by ants, against which he is helpless to defend himself and consequently whose presence highlights the handicap caused by his leprosy: 'He felt an ant sting his back, moved his hand there; the insect moved along his spine; unable to get hold of it he tried to scratch himself, but his soft rubbery fingers failed to obey him. The ant bit him again.'[24] When Baly first meets 'the unhappy leper, alone in his suffering, his smell and the weight of the flies,' he speaks of his battle to overcome his repulsion at the man. His first step towards assisting François is in writing a sign which reads, 'I am a leper, mad and damned,'[25] but he is spurred into further action when, later in the narrative, he is reminded of his lack of charity towards others by Salim. Much to the beggar's surprise, Baly invites him into his home. François is bemused: 'Had he finally found, in this infernal and barren town, an adversary madder than he?'[26] Although the relationship takes time to develop, Baly not only finds a son in François, but eventually also a disciple, an ally, and the future director of his school.

The narrative is replete with descriptions of the ants, fleas, flies, cockroaches and mosquitoes that render the lives of the characters uncomfortable, but insects also become an integral part of some of the characters. François, who is accompanied by his flies throughout the first part of the novel, uses them as a protective barrier between himself and the world, consciously allowing the dirt and flies to accumulate:

> He had tried once and for all to build a barrier of flies and stench between himself and mankind […] It had been enough to let himself go [...] and to live like a plant, slowly, patiently, in a contemptuous solitude, extending his misery like branches, a little more each day, to offend, outrage, shock, horrify, and sully the good conscience of others.[27]

François admits, 'I prefer them to men; [the flies] protect me from their spitting and their blows'[28] and is thrown into a panic when they are chased away: 'they have squashed my companions in misery... they have killed my flies.'[29] The return of the insects is described as a happy reunion, at least on the part of the flies, as 'one by one the flies returned and, buzzing joyfully, landed on his sores.'[30] Insects also appear metaphorically in the novel. Kaba suggests that in Sassine's work, 'The representation of animals […] corresponds in a way to the 'humanisation' of animals and the 'animalisation' of humans. Animals often take on a symbolic character'.[31] In *Saint Monsieur Baly*, Salim the shopkeeper describes the beggars as troublesome insects, remarking that 'they come in greater and greater numbers, like insatiable crickets.'[32] Indeed, when goaded by their parents and the jealous teacher Bana, the school children vandalise the school in a manner akin to the devastation caused by crickets: 'They threw their grass huts into the air, the shrubs were uprooted, the registers and rotas torn up.'[33] The unpredictable nature of these same children is expressed in the description of them as 'odd, slightly mad little animals.'[34] Salim also describes François as 'resembling an insect, or rather a disgusting reptile,'[35] which brings to mind a similar reference in *Wirriyamu*, in which the skin of the albino Condélo is described as 'So cold to the touch [...] It reminds me of the little white wall lizards. I think we call them geckos'.[36]

The proliferation of insects and pests pervades the novel on another level too, interfering with the flow of the narrative. The narrative structure of Sassine's novels is always complex and interesting, and time is often distorted, giving a sense of discontinuity. Jacques Chevrier describes 'the multiplication of points of view, the telescoping of time and of novelistic space, and most of all, the lyricism of the narrative which gives [*Wirriyamu*] the feel of a poetic novel'.[37] In Sassine's later novel *Mémoire d'une peau*, the intricate narrative structure reflects the protagonist Milo's state of mind as he constantly questions his situation and attempts to come to terms with his difference. The multifarious narrative functions on three levels: the novel *Mémoire d'une peau* with Milo as the protagonist, the romantic novel Milo attempts to write, which is also titled *Mémoire d'une peau*, and the letters exchanged between Milo and his lovers at the end of the narrative, which reflect retrospectively on the events we have witnessed as readers.

Similarly disruptive of the traditional novel form, *Saint Monsieur Baly* refers back to the pre-colonial and colonial past, to the difficult period of independence, and forward to the possibilities of a future which, although uncertain, feed Baly's dreams and aspirations. The narrative is constructed around a series of interior monologues, dreams and diary entries. Baly's diary is highly personal and reveals his repeated examination of his conscience, revealing his hopes, aspirations and obsessions. Lengthy extracts are included in the narrative in italics and often mark a transition from intradiegetic to extradiegetic narration. Two central dream sequences are also found in *Saint Monsieur Baly*, both of which occur at significant points in the narrative. The first takes place at the moment where Baly takes the decision to build a school and the second takes place when this same school has been vandalised by those who oppose his project. They occur in periods of crisis, when Baly is faced with practical difficulties or impossible decisions. In addition, Baly's memories interfere with the linearity of the narrative as he returns repeatedly to interrogate past events and encounters. Here, insect imagery is used to symbolise Baly's sense of unease:

> The bug that continued to forage in his memory was right; and all the cadavers that he thought he had buried once and for all, with the sincerity of innocence and pride, seemed to reappear in the triumphant face of his white colleague, and together they cried out the truth to him.[38]

Kaba remarks that 'the pests that take centre stage are harbingers of the agitated and insalubrious atmosphere that gradually pervades the novel'.[39] Indeed, their presence builds and explodes tensions in the novel, as when Baly's wife Fati cries out in frustration 'these flies are crazy!'[40] Baly swats a fly and feels a brief moment of relief before another appears, more obstinate still, reminiscent of the bothersome nature of his memories:

> The fat blue fly filled the air once more with its buzzing, an insidious buzzing, irritating, almost a cry of victory, an invitation to the other flies which steadily returned in groups to the sombre dwelling. [...] Monsieur Baly leapt after it, knocking over once more what his wife had made fall; he followed the guilty fly with a savage joy until he found it under his little bed, and there he slowly squashed it with his thumb, then returned and sat down, released of his anger, sickened.[41]

The insistent presence of the flies, and particularly the sense of nausea evoked here, brings to mind Jean-Paul Sartre's *Les Mouches* (1943), a play with which Sassine would certainly have been familiar. In *Les Mouches*, Orestes' attempts to make a place for himself in the world have obvious parallels with Baly's, and which also examines themes of commitment and responsibility. A rewriting of the Greek myth of Electra and Orestes, *Les Mouches* follows Orestes as he returns to Argos in the company of his tutor to find the city where his father was once King plagued with flies and the people weighed down with guilt. The flies were sent to Argos by the gods fifteen years before the action of the play when Aegistheus and Clytemnestra murdered Agamemnon, and serve as a perpetual reminder of this original sin. The insects in *Saint Monsieur Baly* have a similar function. Baly is deeply affected by his past, by the memories of his forty years as a teacher, which assault him like the big blue fly. The anguish represented by the 'bestiole' or 'bug', prevents Baly from sleeping and bothers his conscience to the extent of a schizophrenic torment: 'Insidiously, a bug started to stir, to plunge into the depths to bring to the surface, somewhat viciously, secrets, numerous small secrets, his little daily cowardly acts, his selfishness...'[42]

Yet, as much as Baly turns his critique on himself and his own actions in the past, *Saint Monsieur Baly* makes clear that others are to blame for the situation against which Baly is struggling in the narrative present. In the wake of colonialism, individuals like Abdoulaye the deputy and Bana have taken advantage to gain wealth and status for themselves. Abdoulaye is

> A corpulent man, with square shoulders, still robust, who admitted to his sixty years of age however when the old lions of independence were spoken of without mention of his name [...] He had two villas in the capital, rented out to ambassadors, as well as lorries, fields of maize and rice in the Niger valley, and friends among the most influential members of the government.[43]

Abdoulaye positions Bana as his man on the inside who will stir up strikes and interfere with Baly's project, planning to 'destroy [Baly] once and for all, he and his ridiculous school.'[44] From the panoramic narrative of *Saint Monsieur Baly*, which exposes the filth, decay, contradictions and conflicts of postcolonial society, emerges a critique of a neo-colonial society. Wealth is in the hands of a privileged minority, which surrounds itself with country houses, cars and consumables that are associated with an acquisitive middle class, while the economic and social position of the underclass is extremely precarious. Those Baly attempts to help experience malnutrition, unemployment, poor health, misery and social alienation. While the novel portrays the underclass as social parasites, the reality is that the privileged minority in power are the most parasitic of all. The insects that pervade the novel are symbolic of widespread corruption, a parasitic force that has eaten and continues to eat into the fabric of postcolonial African society.

Notes

1. A. Sow. 'Forbidden Bodies: Relocation and Empowerment in Williams Sassine's Novels.' *Matutu: Journal for African Culture and Society*. 29–30 (2005), 213.
2. J. Chevrier. *Williams Sassine écrivain de la marginalité*. (Toronto: Gref, 1995), 14. Chevrier notes that Sassine recognised early that his situation was not typical and that this feeling of exclusion left numerous traces in his writing.
3. J. Chevrier and R. Bjornson. 'Williams Sassine.' Research in African Literatures, 23.4 (1992), 133.
4. 'le plus sincère et la plus originale de sa génération'. T. Monénembo, 'La mort d'un juste', Research in African Literatures, 23.4 (1992), 46. My translations throughout.
5. Chevrier and Bjornson, 133.
6. *Ibid.*
7. 'Monsieur Baly, vous êtes usé, vous ne pouvez plus nous servir à grand-chose; votre fonction de conseiller pédagogique est à présent trop lourd pour vous; vos méthodes d'enseignement ont toujours été archaïques [...]; nous vous reprochons surtout de ne pas pouvoir vous entendre avec votre nouveau collègue européen [qui] incarne la coopération avec [la France], et la coopération c'est le nouveau visage de la fraternité dans la francophonie; or vous, vous ne préconisez que la suppression pure et simple du français en faveur de l'enseignement des langues nationales; lorsqu'il vous présenta sa réforme pour une plus rapide diffusion de sa langue, notre langue de culture à tous, vous avez eu cette phrase malheureuse: « L'Afrique n'est pas un terrain d'essai, même pour vos élucubrations pédagogiques: le Sahara devrait vous suffire ». Ça, monsieur Baly, c'est faire de la politique, c'est grave,' 30.
8. For further reading on this subject, see Pius Ngandu Nkashama. *Ecrire à l'infinitif: La dérision de l'écriture dans les romans de Williams Sassine*. (Paris: Harmattan, 2006).
9. Aedín Ní Loingsigh. *Postcolonial Eyes: Intercontinental Travel in Francophone African Literature*. (Liverpool: Liverpool University Press, 2009), 53.
10. 'il y avait également des millions de petits enfants et d'adultes analphabètes qu'il pouvait servir,' 29.
11. 'Arracher son âme aux ténèbres de l'ignorance, à la longue série de mensonges et d'humiliations dont il a hérité des années de colonisation déclarée,' 127–8.
12. 'Le conseiller blanc, le temps et la pauvreté,' 135.
13. 'La joie de vivre que je lis sur les visages des petits écoliers pendant les recréations est ma seule nourriture,' 178.
    14 'Ils constituent la métaphore de la plus grande classe sociale du pays, à savoir celle des pauvres et des déshérités'. Kaba, 31.
15. 'L'image de la grosse mouche bleue et de son abdomen chargé de sang lui brouillèrent un instant la vue. Il se ressaisit rapidement et décida d'avancer vers l'homme-pourri; il n'avait pas encore pitié. Le sentiment qui le poussait vers cette pourriture vivante procédait de l'indentification psychologique et d'une obscure volonté d'illustrer ses vieilles leçons sur la charité'.
16. 'Monsieur Baly et son cortège d'infirmes, de mendiants, de mouches et de microbes représentent eux aussi l'impureté, la souillure de la ville, donc ils doivent être écartés des

lieux saints. Dans la conscience collective des gens de la cité, la chute du vieil instituteur dans l'univers des mendiants l'éloigne des hommes et de Dieu, et le place dans le monde de la marginalité'. Kaba, 39.

17. 'Monsieur Baly, croyez-moi ou non: vous avez une belle lèpre'. Puis il présenta son éprouvette à la lumière et désigna du doigt le sang du vieil instituteur : 'Ce sang a l'air tout à fait normal, mais il est bourré de virus; je les ai vus au microscope,' 104.
18. Kaba, 35.
19. 'Les microbes qui grouillaient dans son sang,' 185.
20. La grosse mouche bleue revint aussitôt à l'attaque, et sans ménagement fondit sur Monsieur Baly, prenant un plaisir particulier à se promener sur son crane fraichement rasé,' 33.
21. Les mouches profitèrent de sa faiblesse pour l'attaquer et bientôt s'agglutinèrent sur son gros crane chauve, ses oreilles et sur chacune des parties découvertes de son corps. Plus il en écrasait, plus il en venait,' 33.
22. 'L'homme–pourri-fou-lépreux-maudit,' 89.
23. 'Alors Toi, mon Dieu, Toi au nom de qui je pratiquais le bien, Tu m'as abandonné et Tu as pris son parti : ma femme et mon enfant moururent peu après de simples maux de tête ; Tu m'as donné la lèpre, regarde mes mains, et mes pieds, et ma face ; je suis tellement pourri que je ne sens plus ma propre odeur ; les créanciers ont saisi tous mes biens et m'ont mis dans la rue. Tous mes concitoyens, du plus petit au plus grand, fuient à ma vue en se signant ; ceux qui osent m'approcher me crachent dessus ou me jettent la pierre,' 14.
24. 'Il sentit une piqûre de fourmi dans le dos, y porta la main; l'insecte se déplaçait à présent le long de son échine; ne pouvant le saisir, il essaya de se gratter, mais ses doigts semblables à des bouts de caoutchouc mou lui obéirent difficilement. Le fourmi le piqua à nouveau,' 16.
25. 'Je suis lépreux, infirme, fou et maudit,' 41.
26. Aurait-il enfin trouvé, dans cette petite ville infernale et désertique, un adversaire plus fou que lui ?,' 90.
27. 'Il avait essayé d'élever définitivement entre lui est les hommes une barrière de mouches et de puanteur [...] Il lui avait suffit de se laisser aller [...] et à vivre comme une plante, lentement, patiemment, dans une méprisante solitude en veillant à étaler seulement ses misères comme des branches, chaque jour un peu plus, pour offenser, indigner, choquer, horrifier, salir les bonnes consciences,' 88–90.
28. 'je les préfère aux hommes; ce sont elles qui me protègent de leurs crachats et de leurs coups,' 86.
29. 'ils ont écrasé mes compagnons de misère... ils ont tué mes mouches...' 65.
30. 'Une à une les mouches le retrouvaient, et dans un grand bourdonnement de joie se posaient sur ses plaies,' 92.
31. 'La représentation du bestiaire [...] correspond en quelque sorte à "l'humanisation" des bêtes et à la "bêtification" des humains. Les animaux [...] prennent souvent un caractère symbolique'. Kaba, 285.

32. 'ils viendront de plus en plus nombreux, comme des criquets, insatiables,' 124.
33. 'Ils firent voler leur paillote de joie, les arbustes furent déracinés, les registres d'appel et les cahiers de roulement déchirés,' 188.
34. 'petits animaux bizarres et un peu fous,' 234.
35. 'semblable à un insecte, plutôt un reptile dégoûtant,' 124.
36. 'Si froid au toucher [...] Ça me rappelle les petits lézards tout blancs des murs, je crois qu'on les appelle geckos'.
37. Wirriyamu est marquée à la fois par la multiplication des points de vue, le télescopage du temps et de l'espace romanesques, et surtout le jaillissement lyrique qui donne à [Wirriyamu] l'aspect d'un roman-poème.'. Chevrier, 144.
38. 'La bestiole qui continuait à fouiner dans sa mémoire avait raison; et tous ces cadavres qu'il croyait avoir enterrés pour toujours, en toute bonne foi de son innocence et de son orgueil semblaient lui apparaitre derrière le visage triomphant de son collègue blanc, et tous ensemble ils lui criaient la vérité,' 58.
39. 'les insectes nuisibles prennent le devant de la scène et constituent un présage de l'atmosphère agitée et insalubre qui envahit peu à peu le roman'. Kaba, 32.
40. 'ces mouches sont folles!,' 53.
41. 33.
42. 'Insidieusement, une bestiole commença à remuer, à plonger pour ramener à la surface, vicieusement, ses secrets, ses innombrables petits secrets, ses petites lâchetés quotidiennes, ses égoïsmes…' 59.
43. 'C'était un homme corpulent, carré d'épaules, encore très solide, qui avouait cependant ses soixante ans avec colère lorsqu'on parlait des vieux lions de l'indépendance sans mentionner son nom […] Il possédait dans la capitale deux villas louées à des ambassadeurs, des camions de transport, des champs de mil et de riz dans la vallée du Niger, et des amis parmi les membres les plus influents du gouvernement,' 121.
44. 'il l'écraserait définitivement, lui et sa ridicule école,' 122.

WORMS POLITIC: PARASITISM, TEXTUAL DECAY AND CONJURE TRUTH IN GLORIA NAYLOR'S *MAMA DAY*

Amy M. Thomas

In Gloria Naylor's *Mama Day*, the climax of the novel centers around one of the main characters, Ophelia, becoming consumed by supernatural worms through a Conjure spell of snakeroot. Yet the narrative tension that unfolds from this parasitic infection is about much more than the fragile line separating life and death, a line which the devouring worms would barely even notice if Ophelia had crossed—it is also about permeating the modern American boundaries between reality and magic, history and myth, truth and uncertainty. In a place where the community's most important histories "don't live in the part of our memory we can use to form words,"[1] the parasites are left as the living tissue of a language of belief. Drawing richly from allusions to Shakespeare and the Western literary cannon alongside symbols from the Conjure practices of African American history,[2] the magical worms and their catalytic implications can be understood as a central metaphor through which these histories infiltrate each other in *Mama Day*.

The novel's interstitial space is first signaled by the landscape itself: the island of Willow Springs exists between states, neither South Carolina nor Georgia, and the people who live there are shaped by enduring but inchoate memories of the woman who brought them out of slavery in 1823. As a direct descendent of this woman, born of the seventh son of her seventh son, Miranda has long been known as a powerful Conjurer and a skilled healer. Like her niece Ophelia, Miranda's name is famously linked to Elizabethan drama, yet she bears little immediate resemblance to her naïve namesake in *The Tempest* (thought the Shakespearean tissue of these names will remain deeply meaningful). When she makes a stand to assure that "nothing parasitic like resorts"[3] will overtake the island as outside developers become interested in their shoreline, no one will go against her word for fear of her Conjure powers: "No point in making a pile of money to be guaranteed the new moon will see you scratching at fleas you don't have." [4] Yet Miranda is also a careful diagnostician and herbalist, impressing the local MD with the soundness of her care practices by biomedical standards. In addition to midwifery, an expertise which contributes to her nickname "Mama Day," Miranda possesses a broad knowledge of natural pharmacology and makes her own medicines, including the warm castor oil and jimson she prescribes for a baby's intestinal worms.[5]

But the real test of her powers does not arise until the end of the novel, when Mama Day's niece Ophelia—whom everyone calls Cocoa—has a powerful spell cast on her by a jealous Conjurer, Ruby. After letting Ruby braid her hair elaborately, like in "prints of African woman with those colored beads draped over their necks,"[6] Ophelia comes back to Mama Day's house and collapses in a fever. Mama Day smells the nightshade in her hair, but she also recognizes that the sickness runs deeper:

> Taking the shears out of her pocket, she begins cutting off each braid. They fall, curled up like worms on the pillow around Cocoa's face. Miranda has to keep her stomach from heaving at the sight of them. She cuts each off close to the head, and with that done she takes the shears and snips carefully through the plaits woven next to the scalp. The white strings pop out, looking even more like worms. Maggots, really. [7]

The spell cast on Cocoa begins with the parasites in her hair, but soon overtakes her whole body as she feels crawling and climbing under her skin. This resonates with one of the most prominent tropes in Conjure practices, the idea of infecting someone with "live things in you." (Indeed, as Yvonne Chireau explains in her essay in this volume, Naylor told Chireau that she grew up hearing such stories as a child in Mississippi.) Through such spells, fantastical parasites become a central vector for causing another person pain or death, as folklorist Harry Middletown Hyatt's informants explain in his voluminous collection of *Hoodoo* accounts: "jes' lak if ah wants tuh *hurt* chew—put a snake in yo…."little somepin like snake worms."[8]

The pulverized dust made from the dried bodies of vermin, bugs, and parasites could also form one of the primary active ingredients in Conjure—goofer dust[9]—which can be used for various spells both of harm and protection. "Yes, any kinda stuff lak dat in *hoodooism* is de *goofer dust,*" another of Hyatt's informants in New Orleans echoed, "…dust from any live insect."[10] Other scholars of *Hoodoo*, including anthropologist Zora Neale Hurston, describe goofer dust as dirt obtained from burial mounds or graves of the dead. While goofer dust (goofer in the sense of goofy, tricksters and tricks) could serve a variety of uses, other practices and recipes were drawn in much more specific terms: little snakes in the human body could be created by making a victim ingest drops of a big snake's blood,[11] or hairs could be bottled in jars and would kill their owners by the time they frayed into snakes.[12] In addition to insects and worms, an astounding diversity of parasites could be conjured through these techniques, as Chireau suggests with her evocative title "Body as Menagerie."[13] These historical Conjure accounts are filled with horrendous and amazing parasitic creatures:[14] turtles in the body of a woman that only move when the moon is full,[15] scorpions that can be wrapped in silk thread in order to make them infect the body of an enemy,[16] or spiders tortured over flames before being singed to dust, resulting in a powder that will reanimate the desiccated creatures internally when drunk by someone in a glass of whiskey.[17]

Given this diverse range of parasitic Conjure spells that could have been worked on Cocoa, it is meaningful that in *Mama Day* Naylor instead evokes only the worm (and the snakeroot that caused it). Profoundly blurring the line between strange reality and fantastical magic, these worms are terrifying because their visceral plausibility hugs the line between madness and death. The mysterious injury that they inflict is deeply internal, almost without physical proof, which makes it all the more unsettling. The parasites' ruinous work on Cocoa's body appears to her only in mirrors, in which she sees her face decaying, the flesh hanging from her chin like "hooked worms."[18] She knows she is battling both the worms and her own encroaching madness:

> There is a limit to how long you can feel your insides being gnawed away without beginning to lose your mind. And I had been fighting to remain sane, for your sake as well as my own. But they were multiplying up toward my throat, and once I saw that I was spitting out worms, it would surely take me over the edge.[19]

Cocoa struggles to discern whether these supernatural worms are physical reality or imagined horrors, and the reader is caught along with her: uncertain to what extent the infestation is chillingly psychic or outright physical, and commensurately unsure how to disentangle Naylor's borrowing from the history of Conjure from her own literary imagination. To some extent, both of these grey areas become meaningfully unanswerable questions in the text, the domain of the worms.

Parasitic worms and small bodily snakes are frequently overlaid in meaning throughout illness narratives around the world,[20] and Conjure healing often shares this intertwined symbolism. In New Orleans, Hyatt heard accounts of the "two-headed" Conjurers from Algiers village, known for the ability to cure people whose bodies where infested with "live snakes." The methods through which these Conjure snakes first enter the body is almost identical to the combing and oiling techniques that Ruby used on Cocoa: "Dey make a dust outa dem an' put dat intuh a Vaseline, or hair oil, or hair grease…an' he passes dat ovah 'is head when he combs 'is hair. An' it [the snake] will penetrate tuh 'is – through 'is skull intuh 'is brain.[21] An' he will always have a [mind] dat he wants to drown 'imself."[22] This last reference—to the fact that the snake induces madness by creating a desire to drown oneself—is particularly striking in the context of Ophelia, whose name is intimately tied to this exact destiny. As Ophelia fights not to drown—in madness, in worms, in water—this Conjure notion recounts a parasitic snake that drives its victims to suicidal drowning, viscerally connects Ophelia's Shakespearean name to the magical parasites, thwarting attempts to separate the signs of either domain. In *Mama Day*, these distinct layers of encoded meanings become interdependent, intertwined and parasitically ingrown. Shakespeare, too, constitutes part of the very tissue—of language and life—that the Conjure worms rearrange.

Parasites of God

Snakes abound in scripture, and Conjure traditions draw powerfully from Christian referents. Yet it is sometimes assumed that snakes inherently denote an archetypal representation of the devil in any biblical tradition, while *Hoodoo* practices in fact often profoundly re-signify such signs, infusing them with layered spiritual meanings (which at times differ dramatically from that of their canonical context). The frequent overlay of worms and snakes further complicates this picture, given that the biological resemblance[23] between the two so often results in intertwined significance.[24] This section samples a few moments in the European theological debate over Christian doctrine and the origins of human worms, in order to contextualize and explore what is specific about the way biblical snakes and bodily worms appear in the American Conjure tradition and in *Mama Day*.

Dating back to medieval debates in the European theological tradition, Christian doctrinal beliefs about mankind's original sin became closely bound with notions of spontaneous generation. Iconic of humanity's putrid matter, worms formed a nexus between these discourses, drawing parasites to the center of a bizarre biblical debate: If Adam and Eve were created in perfection, without parasites, then where did the first human worms come from? Seventeenth-century Italian scientist and naturalist Antonio Vallisnieri argued that not only all future human beings, but also all human parasitic worms were contained in the ovaries of Eve, and therefore must have been transmitted to her in Adam's rib.[25] In *New Apologies of Socrates*, German philosopher Johann August Eberhard argued that parasitic worms *were* present in Paradise, but only as eggs that did not hatch until after the fall of man[26]—insinuating that if the serpent had not been successful in his sinful temptation of Eve, then the parasitic worm eggs within her body would never have hatched. Like miniature cellular recreations of the fall, worms begin eating away at Adam and Eve the very moment they become mortal. Saint Thomas Aquinas also employs a moral version of spontaneous generation in his *Summa Theologiae*, arguing that "the worms which, according to certain passages of the Bible, tortured those in hell, were created there by the putrefaction of their sins."[27]

Such long-standing theological beliefs that link parasitic worms to sin often blended with surprising ease into Enlightenment-age science. In 1792, the learned Society of Copenhagen awarded a gold medal to zoologist Johann August Ephraim Goeze for his twelve proofs that demonstrated intestinal parasites were inborn in the bodies of man,[28] continuing to posit worms as an original flaw built into human biology. Early scientific studies of human worms did little to definitively clarify this philosophical disjuncture with helminth biology. Even after dedicating years to scientific study of worms under a microscope, in 1741 famed French parasitologist Nicolas Andry[29] wrote:

> The worms which are in the human body in the intestines and other parts, in growing old, often take extraordinary forms; some become like frogs, others like scorpions, others like lizards. Some grow horns, others a forked tail, others a kind of beak like the birds, others cover themselves with hairs and become all hairy, others cover themselves with scales and resemble fishes….[T]here is no reason to be astonished that the worms of the body can take after a certain time all these extraordinary shapes which one sometimes observes.[30]

In this proto-scientific view, the parasites' shape-shifting power is physically informed by their host's experiences, so that worms almost become an extension of individual human memory and even the morality embedded in a lifetime of things seen. Andry's descriptions resonate partially with earlier European records of the human "heartworm" causing actions of sin, again linking visceral parasites to the Fall and mankind's vulnerability to temptation.[31] Evolving articulations of Christian theological debates thus became oddly enfolded into early Enlightenment parasitology, through the fundamental notion of worms as inborn flaws. Through an individual's actions, these parasites could be either purified from the body or caused to dangerously return in moments of immorality—in this way continuing to signify man's fundamental imperfection and a direct biological need for God's redemptive power over a lifetime.

In stark contrast to these Genesis-based interpretations of parasites in European theology, the meaning of snakes in the biblical Conjure tradition of *Mama Day* centers on a different book of the Old Testament. As Zora Neale Hurston explains in "The Origins of Voodoo" section of her powerful ethnography *Mules and Men*, it was through a heavenly snake that Moses became the most powerful *Hoodoo* man. Before him, no human understood the magic of God's creation enough to harness his powerful words. "The way we tell it..." Hurston writes of Voodoo and Conjure, such magical words and capacities entered human knowledge because "Moses talked with the snake that lives in a hole right under God's foot rest. ...The snake had told him God's making words."[32] It was this purloined information, imparted by the snake, that helped Moses to free his people from slavery: "He lifted it up and tore a nation out of Pharaoh's side, and Pharaoh couldn't help himself." In this version of the story, the snake is hardly an incarnation of evil from hell or harbinger of mankind's Fall—he exists as an unpredictable third actor in the seeming dichotomy of good and evil. Not a demon, but rather a vermin eavesdropper living in a crack under God's footstool, this snake is a parasite of God.

This dramatic recasting of the snake's primary biblical role is tied to its critical importance in Conjure traditions. Instead of being aligned with the fall of man and original sin, in Hurston's retelling, Moses' snake helps to open a way to freedom for an enslaved people and suggests a margin of imminence and magic—potentially both empowering and destructive—to negotiate oppressive human relations, wiggle room to maneuver between existing frames of good and evil. The bodily worms, little snakes, and other fantastic parasites operating in this system of harming and healing can be understood in light of this moral ambiguity. For example, Hyatt collected several Conjure spells meant to infect a person's body with deadly parasitic snakes that must be accompanied not by pleas to the devil, but instead by prayers to God: "*In de Name of de Fatah, de Son an' de Holy Ghost,* and dey'll become tuh have snakes in em,' or scorpions in em."[33]

Moses' staff famously turned into a snake and ate the courtly staffs of the Pharaoh's magicians. This again underscores the meaning of serpents as they appear in the Conjure tradition, often directly related to wrongly wielded power (a connotation with obvious significance for an American context of slavery and its aftermath, within which *Hoodoo* was at times used against slave owners when no other material recourse was possible[34]—snakes summoned against canes, whips countered with powerful roots, black cats sought[35] to endure cat of nine tails). In Exodus, when Moses grabbed the conjured snake by the tail, it turned back into his staff. This biblical trope resurfaces directly in the methods used to invoke supernatural worms in American Conjure traditions, where blood or dust from a snake's tail are among the most prominent ways to infect a person with bodily parasites. Hyatt reports on the widespread practice of collecting snake blood by severing and draining the serpent's tail during the Civil-War era: "So common was this practice that they say the snakes of Currituck County, North Carolina, have no tails because conjure-men cut all of them off."[36]

This biblical imagery is echoed in *Mama Day* through the carved staff that is handed down from John-Paul to his daughter Miranda, a cane with "the long, sleek bodies of them snakes carved so finely down its length that when he turned it *they seemed to come alive*."[37] This serpent-cane, along with the illegible bill of sale for Sapphira Wade, are the two objects that Cocoa's husband George needs to save his wife from the supernatural worms that are devouring her. In this way, the parasitic worms consuming Cocoa require a connection to the history of Moses' serpent staff (and its parasitic magic) to be effectively fought. At the climax of the novel, this allusion returns again: "Miranda is twirling her walking stick between her knees, watching the carved snakes wind themselves down into the floor and up into her hands."[38] In this description, the snakes of the staff figuratively enter Miranda's healing hands, even as she works to help George save Cocoa from the ravenous worms.

By beginning to tease apart the subtle bricolage of these figures in *Mama Day*, Naylor's use of snakes and worms comes to appear less as easy "syncretism" between Christian referents and Conjure magic than a vital improvisation that emerges from this reshuffling. While European doctrinal debates on worms and snakes linked both figures to mortal sin and binary conceptions of good and evil, these symbols instead come to mark a profound ambiguity in American Conjure practices and in *Mama Day*—worms that can infest a person but also snakes that can liberate them, parasites enabling great harm but also miraculous healing. Neither inherently good nor bad, they have the power to rearrange existing boundaries and realities and are able to either eat away at life or set it free.

Black Hands

When Cocoa falls sick with conjure worms, there is a clear metaphor as George, New York born, becomes anxious to rebuild the bridge that links the island of Willow Springs back to the urban America he knows. He sees Cocoa descending into madness, and although both Doctor Buzzard and Mama Day explain the supernatural roots of her problem to him, their words are not enough to penetrate the rationalist categories through which George has organized his life. He is trying, he repeats again and again, "to hold on to what was real." Mama Day tries to explain to him that she can only heal Cocoa with his help—with his hands, "his very hands"—because his life is inextricably bound up with Cocoa's. (Sometimes love is a parasite too.) But he is unable to process the surreal symbolism of the ancient ledger and the staff carved with snakes that Mama Day tries to hand him. George is not yet willing to accept the overwhelming mysticism of "the Other Place." He returns home to comfort Cocoa in her parasitic sickness, and they make love even in her delirium. That's when it happens.

> As I put up the toilet seat, I thought it was a drop of semen on the end of my penis, but after I urinated it was still there. I used the thumb and forefinger to pry off the clear jellied substance and bring it nearer to my eyes. I looked at it from several angles until I was certain I held a live worm. I smashed it between my fingers. It left a yellowish smear with the odor of rotting garbage. [39]

George holds the worm, examines the parasite and smells its decay, and afterward can no longer ignore the reality of what is happening to Cocoa. In the end, this moment is what incites him to action—not the cajoling of Mama Day or the pain of Cocoa as she rots alive, not even his profound love for her. The final catalyst for George to believe what has happened to Cocoa is the physical, indisputable appearance of the worm itself.

For George, the tangible worm bridges time and meaning that connect the series of seeming dichotomies dividing him and Cocoa: his orphaned childhood opposed to her deep family roots, his meticulous rationalizations versus her outspoken passion, his city upbringing opposed to her ties to a mystical island. The boundaries of any superficial dualities between George and Cocoa break down with the appearance of this tangible parasite: the conjured worm moves between their worlds, and healing must too.

Mama Day and Doctor Buzzard both suggest that the parasitic worms consuming Cocoa's body were put there by Ruby through a spell involving snakeroot.[40] This moment echoes the historical Conjure belief that snake-like bodily parasites can be caused by snakeroots; as one of Hyatt's informants describes, "it was a *root* like a snake in a sack." Hyatt appends his informant's words in brackets: "[this is a *hand*]" the folklorist adds,[41] "hand" being a Conjure term for small "mojo" bags, often flannel sacks filled with magical ingredients that can be used both to harm or to protect.

But instead of countering this Conjure spell with deflective magic, such as a counter fetish or a mojo hand, Miranda instead tells George that what she needs are "his hands in hers—his very hand....So together they could be the bridge for Baby Girl to walk over. Yes, in his very hands he already held the missing piece."[42] Meanwhile, true to her namesake, and shot through with Shakespearean allusion and Conjure magic in equal parts, Ophelia continues to choke on supernatural worms "the same color as the water."[43] Trying desperately to save her, George goes to the chicken coop but can find nothing to bring back for Miranda; he sees only his "gouged and bleeding hands," and even wonders to himself: "Could it be that she wanted no more than my hands?"[44]

George's hands become the dominant motif of his ultimate death that night, overlaying his self-sacrifice with the fetish bag "hands" that connect back to the Conjure origins of Cocoa's illness. Yet the scene also recalls the Gospel of John, where the crucifixion wounds are described "in his hands." (In interviews Naylor describes George as the Christ figure of the book, also dead at age 33).[45] Both Conjure traditions and Christian doctrine are reconciled through the synecdoche of hands, although George's self-sacrifice is not a common motif of Conjure healing, but rather a biblical figure of the New Testament. This notion of redemption shifts the system of exchange. By giving something that was not owed, substituting his own hands in place of the fetish hand, George counters the force of this Conjure spell with love and finds them equals in magic. His untimely death allows Cocoa to survive and the Days to continue. This redemptive figure reveals an important dimension of Naylor's own Christianity around which other spiritual symbols cohere in the novel. George's loving sacrifice regains a future for the days—but for the worms themselves, the implication of his act is only an uncanny equivalence. While before the worms had been eating Cocoa so voraciously that her organs began to turn to dust, George's act healed her at the cost of his own life. Now buried in the family cemetery on the island, he is offered to the worms in her place.

Composition and Decomposition

Once an omen of death throughout much of the American South, a parasite known as the "death watch" beetle lives primarily in wood but also parasitically devours books. It makes a sound like the ticking of a clock. One man in Missouri reported that it especially hovers around people who are ill or otherwise close to death: "Once its ticking stops, the person will die."[46] This recalls literary scholar Daniel Balderston's analysis of parasitic worms eating both textual and human tissue in the work of Augusto Roas Bastos, where the dictator "compares the sound of the clocks with the scratching of the moths in the books."[47] Parasitic imagery connecting time, decay, and textual bodies also appears frequently in the Shakespearean plays that Naylor evokes throughout *Mama Day.* "Let's talk of graves, of worms, of epitaphs; Make dust our paper, and with rainy eyes / Write sorrow on the bosom of the earth" writes Shakespeare in Richard II,[48] linking graves, death and Lady Worm together with the decay of humankind's pages as "dust UNTO dust."[49] As Balderston concludes of the parasitic worms who devour both people and their papers with equal enthusiasm, "The human activities of being concerned with truth and with time, and of becoming involved in fictions, are seen reflected in the worm's [activities], or perhaps each reflects the other."[50]

In this way, the parasite lives between fiction and history, lies and truth, literature and memory. Once committed to paper, any writing becomes a material reality to the parasitic bookworm, who devours each text with equal judiciousness. As philosopher Michel Serres writes in *The Parasite*: "Behind them they leave fragments. Pieces of text and references."[51] In Conjure practices (as in much medieval European magic), scraps of texts often have a major ritual role: Psalms written out of context, old calendars buried in the dirt, victim's names written in invisible ink.[52] In *Mama Day*, these fragmented references encompass both historical figures and literary characters—suggesting names as pieces of magic whose truths are constantly being re-signified, embedded in new stories and fated for altered destinies. Bookworms, as agents of decay and remaking, are thus involved in processes of memory that apply equally to the ethnohistorical recordings of Conjure traditions and to allusions from the Western cannon.

The litany of symbolic names in *Mama Day* is rich and varied, in turns evoking meanings from preexisting contexts or scripts and often working to rewrite them. Miranda lies at the heart of this transformation. In Shakespeare's *The Tempest*, Miranda appears as a subservient daughter to her powerful magician father Prospero. Her offer to carry a bundle of sticks for him, as an equal in work and ability, leads only to his pity: "Poor worm, thou art infected!"[53] But in *Mama Day*, of course, it is Miranda herself who is the island's most powerful magician, weathering a storm explicitly referred to in the text as a "tempest." This is also linked to Cocoa's real name being Ophelia—in a book filled with characters drowning in wells, in water, in history, we are left to wonder whether she will drown too in this dark spell of conjured worms that coincides with a storm borne from the Atlantic and the Middle Passage. But like Miranda, Naylor's Ophelia is stronger and more resilient than her Shakespearean namesake. She survives the waters of the storm.

Likewise, the figure of Sapphira Wade echoes Willa Cather's novel *Sapphira and the Slave Girl*,[54] with a significant recasting: instead of "the slave girl" serving as a supporting character for the ill-intentioned white healer Sapphira, in Willow Springs Sapphira Wade is center of her own story and social universe, a slave woman who frees herself and all of her descendents on the island. Through this act, Sapphira becomes something of a female remaking of John[55] the Conqueror, a prince who in Conjure tradition freed himself from slavery and returned to Africa, leaving power behind in his roots.[56] The novel's reordering of these historical and literary figures bears an intimate tie to the parasites in Shakespeare's *Hamlet* (a text which is also referenced in *Mama Day*), particularly the "certain convocation of politic worms" that feast with diplomatic evenness on the bodies of both emperors and beggars.[57] Through the worms, Hamlet later muses, dust from the corpse of Alexander the Great could be used to stop up the hole in a beer barrel; any reordering at all could take place through the parasitic processes of decomposition and remaking, just as the fate of the borrowed characters in Naylor's own text are reworked from their historical constraints.

Mama Day also draws implicitly from Charles Chestnutt's 1899 collection *Conjure Woman*, as literary scholar Lindsey Tucker describes.[58] Yet Chestnutt himself was also drawing from an array of fragmented signifiers floating between folklore and fiction, at times even evoking them unknowingly. For example, the final tale in his collection of seven stories is 'Hot Footed Hannibal,' in which the Conjure woman fashions a fetish doll with red pepper pods for feet.[59] Chesnutt later admitted that at the time of writing he thought the red peppers "a purely original creation," until he later learned this mirrored an actual Conjure practice of fetish "Hotfoot powder."[60] As Chesnutt described of the work, "I discovered that [some of] the brilliant touches due, I had thought to my own imagination, were after all but dormant ideas…" He concluded that "creative talent" was in actuality "the power of rearrangement"[61] of existing truths, intuited or re-invoked rather than invented.[62]

This process of intuiting missing truths is much more than a problem of gaps in written history; throughout *Mama Day*, books and written documents cannot be trusted. The fact that texts can violently misrepresent people as often as they capture truth is set up as a fundamental irony from the novel's very beginning: Sapphira Wade's bill of sale, which describes her as a "half prime," constitutes the first page of text in the book. By the time Miranda finds her great-grandmother's Bill of Sale in the attic, the ancient "crumbling paper"[63] has been pulverized almost completely to dust. Only a few random words and the letters *Sa* are still legible. Yet Miranda is able to attune herself to this history, and in her dreams she comes to know Sapphira by name.

In this way, the untrustworthiness of written texts in *Mama Day* is the opposite of memory's decay—instead this unreliability of writing demands deeper channels of connection to access time and history. All tissue of language and human history will one day be deconstructed by the worms. Yet the very holes that parasites leave in knowledge are part of what keeps Miranda porous to the traces of history and memory that may never even enter language but can still be read through dreams, conjure, *listening*.

Similarly, the novel opens with a book of untrue facts being composed by an anthropologist, a premise that is both biting and deeply funny.[64] The island community of Willow Springs is being studied for the first time by an ethnographer, born in the same community where he later researches his book. He is identified only as Reema's boy, "the one with the pear-shaped head." The anthropologist wants to study the expression 18 & 23, a topic that could have led a thoughtful listener to clues about the heart of Sapphira Wade's story and the freedom she won in 1823. But instead, the book he writes captures none of the truth about 18 & 23 or its meaning:

> ...he sent everybody he talked to copies of the book he wrote, bound all nice with our name and his written on the front page. We couldn't hold Reema down, she was so proud. It's a good thing she didn't read it.... you see, after "extensive fieldwork"...he made it to the conclusion that 18 & 23 isn't 18 & 23 at all—was really 81 & 32, which just so happened to be the lines of longitude and latitude marking off where Willow Springs sits on the map...called it "asserting our cultural identity," "inverting hostile social and political parameters." 'Cause, see, being we was brought here as slaves, we had no choice but to look at everything upside-down."[65]

The anthropologist's invented interpretation of facts echoes Baldersten's observation that "historians who are true to the letter but falsify the spirit of an event gnaw away at the truth of documents as surely as do bookworms and rats."[66] Reema's son invents a poor theory (misconstrued from a cartographic record) in an effort to fill in a gap in conscious history, unable to understand the importance of unspoken truths. But the voice of the island insists on truths beyond language: "[L]isten. Really Listen this time....you done heard it without a single living soul really saying a word."[67]

This anthropologist's failed imperative to correctly interpret signs reveals the ethnographer's craft, too, as a form of conjure—sensed truths, intuited relations, feeling one's way toward understanding. Yet ethnographic texts also decay and parasites return, just as Sapphira's Bill of Sale has crumbled to illegible dust yet still remains important for George to carry in warding off Cocoa's worms. Even within a relatively recent text—such as the 1978 folklore compendium by Harry Middleton Hyatt used in this essay—there are glaring holes in the transcribed record, important sentences interrupted by underlined blank spots and bracketed references to "bad static." And static, as Michel Serres reminds us, is built into the French definition of the word *parasite*—acoustic worms, white noise.

It is around one of these blank spots in the text that an informant tells Hyatt of oral histories recalling "Doctor Buzzard and the Seven Sisters." The actual Gullah historical figure of Dr. Buzzard (documented by Chireau and other historians of religion)[68] surfaces in *Mama Day* as an old charlatan by the same name, another intertextual fragment. While transcribing, Hyatt notes of the interview where Doctor Buzzard and the Seven Sisters are mentioned that "[Static is bad here],"[69] and only a brief description of Doctor Buzzard follows in his text. The story of the Seven Sisters, who are alluded to as strong Conjurers and healers, is ultimately never told—there is simply a blank spot in the recording, white noise. As the seventh child of a seventh child, *Mama Day*'s Miranda in some ways conjures—through fiction and memory—the historical account of the Seventh Sister that is missing from written history in the static-filled record of Hyatt's fieldwork encounters.

Naylor's own divination process is captured by the novel, but its conjuring arts also transcend the written text which will one day be subject to the worms. This is suggested as we are first introduced to Miranda, "humming—humming some lost and ancient song."[70] The island of Willow Springs answers back: "gators could sing,"[71] "morning glories started to sing,"[72] in fact all "flowers can be made to sing."[73] This music is both written into Naylor's text and remains alive in collective memory, unlike the riddle of the bookworm as rendered by Borges: "A worm eats words…the worm, a thief in the darkness, has devoured the famous song of man and its fundamental strength." As Balderston analyzes the original Saxon riddle this translation derives from: "The riddle is a reflection on the fragility of written literature, subject as can be seen to the depredations of ignorant insects, and is implicitly a celebration of the relative permanence of oral literature."[74]

This implicit theory of parasites has deep meaning for understanding the material and intangible records of African American history—indeed, all history. In fact, the worms take any assumptions about oral traditions being ephemeral while written ones are permanent and eats them away too.

As a novel written in three parts, between two lovers (one of them dead) and an island, *Mama Day* itself reads like an ultimate act of dialogue—a call and response memory that makes itself heard through the parasitic static of history. When the tempest comes, there is a whispering of a wind that must be Sapphira—"*I call to remembrance my song in the night*"—and "Miranda hears it in her soul."[75] The musicality to the dialectic structure of *Mama Day*, then, should come as no surprise—even Serres, for whom all relations harbored worms, believed that musical *accord* was the only way to overcome the parasite's noise, eliminate its static. In a medical extension of this philosophy, Reinhard Hoeppli describes reports throughout Europe noting that tapeworms find music profoundly disturbing, obligating their hosts to run away from its source.[76] Goeze further elaborates on their acoustic preferences: "I have noticed fear and disagreeable sensations regarding music in more than one case of *taenia* [tapeworm] infection. On this account the sound of the organ makes them feel unwell at church. This reminds me of the statement by Lange(n) in Lüneburg, that one could expel worms in man by the noise of the Jew's harp."[77]

Perhaps it is no wonder that these politic worms—simultaneously human parasites and static hisses in the record, devourers of the written word and catalysts of change—are opposed to collective music, which is valuable in *Mama Day* precisely because it fixes something precious in collective memory, offers a way to call out to lost history while acknowledging the memories that cannot be brought into language. Again: "Uh, huh, listen. *Really listen this time*: the only voice is your own. But you just done heard about the legend of Sapphira Wade, though nobody here breathes her name."[78]

Besides being a novel about the love of Conjure women and the specter of worms (that is, the specter of death itself), reading Naylor's *Mama Day* fundamentally evokes a sense of conjuring ghosts from a sea of time or recalling someone else's memory—not a divide between oral and written traditions, Western literary cannons or Conjure histories, memories either sensed or spoken—but the reinforcing power of intuited truths and tunes that occasionally, miraculously, escape the reach of the worm. History forgotten but still somehow viscerally known, this memory lives on in *Mama Day,* as people in Willow Springs link arms at Candle Walk and together "hum some lost and ancient song."[79] Or, in Serres' words:[80] "Together. A vibration of several voices. Coming together. The collective... music has spoken in tongues. It is free of parasites."

Acknowledgements

I am grateful for the American University Student Research Grant that supported this archival research in its beginnings, with special gratitude to the memory of Betty Bennett. Thanks also to Richard Sha, Roberta Rubenstein, Jim Boon, João Biehl, Carol Greenhouse, Carolyn Rouse, Franco Rossi, and my parents for their contributions of time and care throughout various stages of this research.

Notes

1. Gloria Naylor, *Mama Day* (New York: Vintage Press, 1988), 4.
2. In my readings of Conjure traditions and their literary evocations I follow Yvonne Chireau, whose delves into the *Hoodoo* practices of enslaved black Americans and their free descendants as difficult, partial yet revealing windows into the embodied ethics and spiritual worlds of people reckoning with the visceral realities (and more recently, spectral histories) of slavery and its violence. See Yvonne Chireau, *Black Magic: Religion and the African American Conjuring Tradition* (Berkley: University of California Press, 2003). Chireau's chapter in this volume gives crucial historical context for my own essay on the literary implications of such parasite tropes as they appear in Naylor's *Mama Day*. For other work that examines issues of subjectivity, agency and bodily experiences of racial oppression as encoded in Conjure practices, see also Jason Young, *Rituals of Resistance: African Atlantic Religion in Kongo and the Lowcountry South in the Era of Slavery* (Baton Rouge: Louisiana State University Press, 2007); Jeffrey E. Anderson, *Conjure in African American Society* (Baton Rouge: Louisiana State University Press, 2005); Sharla Fett, *Working Cures: Healing, Health and Power on Southern Slave Plantations* (Chapel Hill: University of North Carolina Press, 2002); Walter Rucker, "Conjure, Magic and Power: Influence of Afro-Atlantic Religious Practices on Slave Resistance and Rebellion," *Journal of Black Studies* 32:1, 84–103 (2001); and Theophus Smith, *Conjuring Culture: Biblical Formations of Black America* (Oxford: Oxford University Press, 1994). When discussing the modern boundaries inscribed between reality and magic, I begin here from the premise that these divisions today are not 'natural' but themselves the product of particular formulations and legal rationalities of the Western nation-state. In alternate frames of spiritual and juridical reference such as Conjure, the modern biotechnical categories of magic, religion and science instead operate along meaningful gradients, often symbiotically overlaid, functioning with their own specific rationality and logics of accounting. For explications of this theoretical stance in other global milieu, see, for example, Edward Evans-Pritchard, *Witchcraft, Oracles and Magic Among the Azande* (Oxford: Clarendon Press, 1937); Richard Kieckhefer, *Magic in the Middle Ages* (Cambridge: Cambridge University Press, 1989); and Carol Greenhouse, *A Moment's Notice: Time Politics Across Cultures* (Ithaca, NY: Cornell University Press, 1990).
3. Naylor, *Mama Day,* 185.
4. Naylor, *Mama Day,* 6.
5. Naylor, *Mama Day,* 192.
6. Naylor, *Mama Day,* 247.
7. Naylor, *Mama Day,* 264.
8. Harry Middleton Hyatt. *Hoodoo-Conjuration-Witchcraft-Rootwork* (Cambridge, MD: Memoirs of the Alma Egan Hyatt Foundation, v.1–5, 1978), 227.
9. Goofer dust is alternately documented by other scholars of Hoodoo (including Zora Neale Hurston) as dirt gathered from burial mounds or graveyards, rather than desiccated

insects (as in this later variation that Hyatt records in South Carolina and Georgia). Goofer dust as linked to graveyard dirt was perhaps more closely tied to African associations, as Jason Young suggests: "the term 'goofer' derives from the kiKongo *kufwa*, meaning dead person." Young, *Rituals of Resistance,* 163. This older etymology seems to have become overlaid with the English resonance of the word "goofer" by the time Hyatt records this association between goofer dust and being "goofy," acting a trickster. For more on the literary implications of this trickster connotation (which echo throughout this essay), see Henry Louis Gates, *The Signifying Monkey: A Theory of African American Literary Criticism* (Oxford: OUP, 1988).

10. Hyatt, *Hoodoo,* 226.
11. Hyatt, *Hoodoo,* 231.
12. Hyatt, *Hoodoo,* 3103.
13. Yvonne Chireau, "Body as Menagerie: Sickness and African American Hoodoo Narratives," *Bulletin of the Council of Societies for the Study of Religion* (2007).
14. Yvonne Chireau, "Invasion as Affliction: Worms and Bodily Infestation in African American *Hoodoo* Practices" (this volume).
15. Hyatt, *Hoodoo,* 3106.
16. Hyatt, *Hoodoo,* 3087.
17. Hyatt, *Hoodoo,* 3103–4.
18. Naylor, *Mama Day,* 276.
19. Naylor, *Mama Day,* 290.
20. In addition to the other chapters of this volume, see EC Green, "Purity, Pollution and the Invisible Snake in Southern Africa," *Medical Anthropology* 17: 83–100 (1997); David Grove, *A History of Human Helminthology* (Oxon, UK: CAB International, 1994); Nobert Vecchiato, "Digestive Worms: Ethnomedical Approaches to Intestinal Parasitism in Southern Ethiopia," in *The Anthropology of Infectious Disease* (Amsterdam: Gordon and Breach Publishers, 1997); and Wenzel Geissler, "Worms Are Our Life: Understandings of Worms and the Body Among the Luo of Western Kenya," *Anthropology & Medicine* 5:1–2 (1998).
21. Just as Miranda intuited the need to cut off Cocoa's wormy braids, the Conjurers of Algiers also try to prevent death from this parasite by trying to "shave all de hair off." Hyatt, *Hoodoo,* 3091.
22. Hyatt, *Hoodoo,* 3090.
23. "A large specimen of *Ascaris lumbricoides* vomited out or discharged in the stool could easily be mistaken for a snake as there was no clear zoological differentiation between roundworms and snakes." Reinhard Hoeppli, *Parasites and Parasitic Infection in Early Medicine and Science* (Singapore: University of Malaya Press, 1959), 92.
24. See Brenda Gardenour, this volume, as well as Alison More, this volume.
25. Hoeppli, *Parasites and Parasitic Infection,* 97; cf Arthur William Meyer, *The Rise of Embryology,* (California and London: Stanford University, 1939), 66.

26. Hoeppli, *Parasites and Parasitic Infection,* 401; cf Rudolf Leuckart, *Die Parasiten des Menschen und die von ihnen herruhrenden Krankheiten* (Leipzig and Heidlberg: CF Winter, 1886), 35.
27. Hoeppli, *Parasites and Parasitic Infection,* 84; cf Edmund von Lippman, *Urzeugung und Lebenskraft*, (Berlin: Julius Springer, 1933), 33.
28. Hoeppli, *Parasites and Parasitic Infection*, 98.
29. Andry also famously described human spermatozoa as "spermatic worms," linking parasites to the most integral genetic material of humanity, in what could be considered a scientific reformulation of the idea that the first worms came from Adam's rib.
30. Hoeppli, *Parasites and Parasitic Infection,* 99–100, cf Nicholas Andry, *De la Génération des Vers dans le Corps de l'Homme* (Paris, 1741), 281–4.
31. Peter Dendle, *Satan Unbound: The Devil in Old English Narrative Literature* (Toronto: University of Toronto Press, 2001), 29.
32. Zora Neale Hurston. *Mules and Men,* (New York: Harper Collins, 1990 [1935]), 184.
33. Hyatt, *Hoodoo,* 3094.
34. Hyatt, *Hoodoo,* 242–3.
35. In Conjure traditions, black cat bones are a notorious sign of a pact, usually made with the Devil, for supernatural powers (Hurston 220), a symbol made famous in the twentieth century United States by rumors surrounding Robert Johnson and other legendary blues musicians. In *Mama Day*, only the charlatan Doctor Buzzard wears black cat bones around his neck (Naylor 163, 292).
36. Hyatt, *Hoodoo,* 231.
37. Naylor, *Mama Day,* 88 (emphasis mine).
38. Naylor, *Mama Day,* 266.
39. Naylor, *Mama Day,* 298.
40. Naylor, *Mama Day,* 287. For a broad background on the principle of sympathetic magic throughout other historical and cultural contexts, see Sir James George Frazer, *The Golden Bough* (New York: Macmillan Limited Editions, 1922).
41. Hyatt, *Hoodoo,* 242.
42. Naylor, *Mama Day,* 285.
43. Naylor, *Mama Day,* 297.
44. Naylor, *Mama Day,* 300.
45. Sharon Felton and Michelle C. Loris, in *Conversations with Gloria Naylor,* (Jackson, MS: University of Mississippi Press, 2004), 143.
46. Niles Newbell Puckett, *Folk Beliefs of the Southern Negro* (Chapel Hill, NC: University of North Carolina Press, 1926), 493.
47. Daniel Balderston, "Eater-Reception and De-Composition: Worms in *Yo El Supremo,*" *Modern Language Notes* 101:2 (1986): 420.
48. William Shakespeare, *Richard II* (New York: Simon & Schuster, 1996 [c. 1591]).
49. The concept of dust onto dust is also linked to worms specifically in this Conjure healing account: "One winter night when [an old slave conjure-doctor who is now dead] was

racked with rheumatism, an angel appeared to him in a vision informing him that man was made of dust and that it would take dust to patch him up when he was ailing. *Earthworms are the life of dust*, therefore fry the earthworms in lard and use the mixture as a salve." Niles, *Folk Beliefs*, 362–3.

50. Balderston, "Eater-Reception and De-Composition," 420.
51. Michel Serres, *The Parasite* (Minneapolis: University of Minnesota Press, 1980), 237.
52. Hyatt, *Hoodoo*, 4572.
53. William Shakespeare, *The Tempest* (New York: Simon & Schuster, 2004 [c.1611]), 37, line 32.
54. Willa Cather, *Sapphira and the Slave Girl* (New York: Vintage, 1975 [1940]).
55. John is also the birth name of Miranda's father in the novel.
56. Zora Neale Hurston, *The Sanctified Church* (New York: Avalon Marlowe & Co, 1988).
57. William Shakespeare, *Hamlet* (New York: Simon & Schuster New Folger, 2003 [c. 1601]), Act IV, Scene III.
58. Lindsey Tucker, "Recovering the Conjure Woman: Texts and Contexts in Gloria Naylor's Mama Day," *African American Review* 28:2 (1994): 178–88.
59. Charles Chesnutt, *The Conjure Woman* (Boston & New York: Houghton, Mifflin and Company, 1899).
60. Robert Hemenway, "The Functions of Folklore in Charles Chesnutt's 'The Conjure Woman" *Journal of the Folklore Institute* 13:3 (1976): 283–309, 293.
61. Hemenway, "Folklore in Charles Chesnutt's 'The Conjure Woman,'" 286.
62. Chesnutt's reflections here resonate powerfully with Henry Louis Gate's later analysis of *Their Eyes Were Watching God.* As Gates writes in his seminal work on the theory of Re-Signification: "It is Hurston's sense of revision, defined as "originality [in] the modification of ideas," and "of language" as well as "reinterpretation," which I have defined…as the ultimate meaning of the trope of Signifying….This notion of imitation, repetition, and revision, she maintains, is fundamental to "all art," indeed is the nature of art itself, even Shakespeare's." Henry Louis Gates, "Zora Neale Hurston and the Speakerly Text," in *Their Eyes Were Watching God: A Casebook* (Oxford, OUP: 2000), 92.
63. Naylor, *Mama Day*, 280.
64. See Robin Blyn, "The Ethnographer's Story: 'Mama Day' and the Specter of Relativism," *Twentieth Century Literature* 48:3 (2002): 239–263.
65. Naylor, *Mama Day*, 7–8.
66. Balderston, "Eater-Reception and De-Composition," 211.
67. Naylor, *Mama Day*, 10.
68. Chireau, *Black Magic* 24–7.
69. Hyatt, *Hoodoo*, 4513.
70. Naylor, *Mama Day*, 4.
71. Naylor, *Mama Day*, 205.
72. Naylor, *Mama Day*, 152.

73. Naylor, *Mama Day*, 139.
74. Balderston, "Eater-Reception and De-Composition," 418.
75. Naylor, *Mama Day*, 250.
76. Hoeppli, *Parasites and Parasitic Infection,* 108.
77. Hoeppli, *Parasites and Parasitic Infection,* 108-9.
78. Naylor, *Mama Day*, 10.
79. Naylor, *Mama Day*, 111.
80. Serres, *The Parasite*, 134.

INVASION AS AFFLICTION: WORMS AND BODILY INFESTATION IN AFRICAN AMERICAN *HOODOO* PRACTICES

Yvonne Chireau

> When the medical doctor does you no good, you should go to the Conjure doctor. He will give you a brew made out of roots and herbs…then will …come out little ground puppies…they are given this name because they have heads and tails like little dogs and they are always wiggling.[1]

Descriptions of worms and worm-like animals like salamanders, snakes, and tadpoles are a remarkable feature of African American affliction narratives in late 19th and early 20th centuries. Often appearing as unwelcome internal inhabitants of the human body, the creatures are envisioned as the material manifestations of illness. However, the fact that many descriptions of bodily infestation derive from second-hand testimonies, unconfirmed eyewitnesses, and hearsay reports has led to their dismissal as a kind of "folklore," "delusions" or simply "superstitions." On the other hand, one might also argue that these descriptions of bodily infestation possess their own integrity as historical documents, in that they present new possibilities for interpreting embodied experience. Furthermore, because many of the accounts are embedded within discourses of spirituality and religion, they are especially interesting for what they suggest about worldviews within given populations. Beyond their use as texts for theorizing sickness, stories of infestation also provide valuable evidence of the ways that African Americans have historically constructed and mapped their inner worlds. In this essay, I will examine accounts of worms and animal infestations that appear in African American *Hoodoo* narratives. *Hoodoo* is a spiritually-based healing and harming practice that dates back to slavery. In their conceptualizations of affliction within the *Hoodoo* tradition, African Americans simultaneously imposed and negated boundaries of the self and provided a rich, densely organized subjectivity by which they conceptualized and articulated physical suffering. A closer investigation of African American narratives of *Hoodoo* infestation reveals a vast and colorful complex of ideas and expressions that have effectively interacted with the external social realities of African American life.

Descriptions

> ...it was no illusion that they had begun to crawl within my body. I didn't need a mirror to feel the slight itching as the curled and stretched themselves, multiplying as they burrowed deeper into my flesh...It was reaching a point where I couldn't stand it anymore. There is a limit to how long you can feel your insides being gnawed away without beginning to lose your mind...but they were multiplying up toward my throat, and once I saw that I was spitting out worms, it would surely take me over the edge.[2]

This excerpt from Gloria Naylor's 1989 novel, *Mama Day*, reveals the mental and physical agony of worm infestation from the imagined point of view of a victim. Naylor told me that she had grown up hearing about such things when she was a child in Mississippi, where folk traditions and stranger-than-fiction storytelling were important sources of vernacular knowledge, as they were in black communities throughout the South. Naylor's created character is unique because she provides a self-described experience of bodily infestation, whereas most narratives of *Hoodoo* affliction derive from second- and third-person perspectives, and are not normally offered by victims. This is significant, because while African American *Hoodoo* narratives of infestation utilized distinctive imagery to articulate physical distress, they did not privilege the voice of the sufferer in doing so. While the sick and afflicted themselves were unlikely to describe their bodily symptoms as infestation, other observers, such as family members, neighbors, and friends, constructed graphic accounts, creating a profoundly empathetic documentary of illness.[3]

I believe that *Hoodoo* stories encode culturally-specific ways of depicting disorder and disease for African Americans. Recurring descriptions of creatures circulating throughout the bodily system, contained but dangerous, can be read as signs of distress. These are social narratives in that they link personal affliction to community concerns and relationships, taking take full consideration of the environment and its actors in the telling of the tale. In this report, for example, sickness is narrated as a local affair by an unidentified spectator, yet it implicates specific places, persons, and designated roles of patient and healer in the infestation plot:

> In 1873 I was going down the street and came across a Conjure doctor, and he asked me where did a certain lady live, and I told him and he went there; and when he got

> there the woman told him she had some kind of pain in her head and side, and told him something kept coming up in her throat and she did not know what to do...[4]

This account outlines a typical *Hoodoo* story, literally placing the affliction and its treatment within a context of treatment. According to the narrator, the "Conjure" doctor gives the sick woman his own medicinal formula, and "in five minutes," he notes "a scorpion came out of her mouth." This narrative, like so many others, details a specific illness process that was well known to a significant segment of African American people in the United States from the nineteenth century onward. As they are told, *Hoodoo* stories appear to be fantastic, but they are nearly always offered with the sober and straightforward delivery of blunt truthfulness.[5]

Observers corroborate the facts of bodily infestation. Ex-slave Nathaniel John Lewis of Tin City, Georgia, declared that he had been witness to an infestation of serpents in his wife's body. "My wife Hattie had a spell put on her for three long years with a nest of rattlesnakes inside her," he insisted. "She just lay there and swelled and suffered." Other reports of *Hoodoo* related afflictions describe the invasion of the body by worms and worm-like amphibians such as snails, salamanders and tadpoles. In the early 1900s, sociologist Newbell Niles Puckett spoke to an ex-slave in Mississippi who suspected that he had been "tricked" by an enemy who put poison into a glass of whiskey. "He held the glass in his right hand and in a short time that had began to swell," stated Puckett. "Great was his astonishment, on slitting open the palm of his hand with a small knife, to find that the inside...was alive with small black-headed worms." Similar depictions, related in graphic detail, create a visceral picture of the pain and horror that was believed to be experienced by victims of bodily infestation.[6]

Accounts of *Hoodoo* sickness routinely testified to plagues of snakes, lizards, and other creatures in the body. Former slave George Little, a "root" doctor skilled in domestic medicine from Brownsville, Georgia, observed the deadly consequences of such infestations. "Frawgs an lizuds and sech tings is injected intuh people's bodies an duh people den fall ill an sometime die," he said. Such matter-of -fact statements were always relayed with the firm conviction of their truth. "...since I came to talk with such good old Christians who I know would not tell a lie," declared one writer in a late-nineteenth century account, "...they told me they have seen ground puppies (salamanders) jump out of folks feet, and lizards and snakes." Another eyewitness stated that he had viewed a *Hoodoo* doctor who had cured a patient by taking "out of her right arm a spool of thread, and out of her leg a lizard...this is the truth," the witness professed, "what I saw with my own eyes." Ex-slave Estella Jones of Augusta, Georgia, also insisted on the veracity of her version of events. "Some folk don't believe me, but I ain't telling no tale about it," she maintained. Jones said she had been an eyewitness to several cases of *Hoodoo* poisoning, and admitted that it was easy for other persons "to put snakes, frogs, turtles, spiders, or most anythin' that you couldn't live with, crawlin' and eatin' on the inside of you." Finally, in a 1940 Federal Writers Project interview, another Georgia resident described a relative that had been poisoned while drinking alcohol with a spiteful and jealous neighbor. After he got sick, she said, "He told everybody that asked him...about the frogs in his stomach:"

> The bigger these frogs got, the weaker he got. After he had been sick about four months and the frogs had got to be pretty good size, you could hear them holler every time he opened his mouth. He got to the place where he wouldn't talk much on account of this. His stomach stuck out so far, he looked like he weighed 250 pounds. After these frogs started hollering in him, he lived about three weeks, and before he died you could see the frogs jumping about in him and you could even feel 'em.[7]

I have found accounts of *Hoodoo* infestation to be peculiar to the illness narratives of blacks in the United States. *Hoodoo* affliction first came to the attention of writers in the post-Emancipation period, when the former slaves, their sons, and their daughters began to reflect publicly on the lives that they lived. Newspaper articles, oral histories, autobiographies, and short story collections drew upon the words of African American informants and sometimes circulated as interest pieces in the popular press, occasionally providing unique data for social scientific studies. *Hoodoo* narratives (also known as Conjure tales), emerged as a mixed genre of African American literature, as they included both oral and written sources, fictional and non-fiction texts, and vernacular and academic prose. One of the earliest bodies of collected *Hoodoo* illness narratives, for instance, was a compilation of letters submitted to the school newspaper at Virginia's Hampton Institute in the late 1870s. African American student members of the Hampton Folklore Society gathered the information for a folklore and ethnology project that was believed to possess both scientific and social value. An excerpt from the following essay, which is adopted almost verbatim from the student papers, describes African American *Hoodoo* practices and beliefs with great trepidation. To be sure, the co-authors of the piece, two white women of the northern liberal Protestant establishment who had become teachers at the freedmen's college, viewed *Hoodoo* as a "backward" and "primitive" survival practice from the descendents of African heathens. In the article they document the processes by which victims believed the human person to be afflicted by *Hoodoo* with the intrusion of animal entities in the body:

> The disease…may be recognized in its early stages, in the first place the suddenness of the attack. The victim is seized with a sharp pain in some part of her body; later, swelling and other symptoms follow, but the beginning of the attack can usually be traced to a pain which followed directly upon handling, stepping over, or swallowing a charm...As the disease develops itself the symptoms become more severe and terrible in their nature. In many cases snakes and lizards are seen running up and down under the flesh, or are even known to show their heads from the sufferer's mouth. One example is given of a woman possessed by a lizard that "would run up and down her throat and hollow (holler) when she would be a talking. Another case is of a man whose food did him no good. The conjure doctor told him that he had been conjured and that inside of him were a number of small snakes which ate up the food as fast as he ate it. Another woman who had lizards crawling in her body was obliged to eat very often to keep the lizards from eating her. This possession by reptiles of various kinds seems to be a part in almost every evil wrought by the conjurer, and instances are too numerous and too horrible for a more detailed review of them in this paper.[8]

The detached, almost clinical style of presentation here contrasts with the personal *Hoodoo* stories that were told and retold by black Americans in the early twentieth century, although the structure of the accounts is the same: an individual, clearly afflicted by some inexplicable ailment, complains of great pain while experiencing other symptoms, such as the inability to eat or drink. Then, some unpredictable physical disability occurs. The acute distress of the afflicted person– who is nearly always kin, a neighbor, or a member of a local community – is made real as their pain is rendered in spectacular form by concerned observers. To the reader, many of these illness accounts display a kind of visual empathy, in that they communicate the immediacy of a victim's suffering as well as an awareness of the larger context in which the crisis emerges.[9]

Hoodoo

What is *Hoodoo*? *Hoodoo* is a complex of ideas and actions that originated among enslaved African Americans for purposes of personal empowerment, healing, and physical protection, as well as retaliation, aggression, and defense against perceived oppressors. *Hoodoo* uses symbols, rituals, and language adopted from Christianity, indigenous African faiths, and other cultural sources, including Native American and European traditions. As a practice with religious referents, *Hoodoo* mediated spirituality and embodiment through the enactment of harming *and* healing – that is to say, *Hoodoo* was believed to both afflict and to cure affliction. Also known as Conjure, *Hoodoo* was widespread among slave populations in the United States during the 1800s, and later among freedpersons. With *Hoodoo*, some African Americans believed themselves able to manipulate spiritual forces and influence the destiny of other human beings.[10]

Hoodoo sicknesses always produced visible side effects, the most unusual of which was bodily infestation. The incursion of animals within the body – small, teeming creatures of a lower order - is viewed in African American sickness narratives as clear evidence of *Hoodoo* affliction. The effects of supernatural harming are seen in the painful physical symptoms and the visible "proof" of the intruding presences, worm-like entities that move about without restriction within the body of the afflicted individual. But how was it thought that these invasions of the body came about? Black Americans believed that *Hoodoo* brought harm to victims by causing misfortune, sickness, or death with supernatural maleficia. Practitioners utilized a variety of measures in order to afflict, including ritually-empowered artifacts and objects, elaborate oaths, curses, and prayers – these were common practices in *Hoodoo* traditions. However, bodily infestations were most often believed to be triggered by one of two specific methods: they were either caused by substances that were activated by eating or drinking them, or they were brought about through forces that required indirect contact with the intended victim. "A bottle in which a lizard has been put is placed on the road," states one source, "…and if the unfortunate victim shall step over that bottle…the lizard miraculously hops into him, and he ends his days in agony." One of the most well known methods of *Hoodoo* poisoning involved a kind of sympathetic magic that was brought about by ingestion. One witness described an African American man that had been made ill, he believed, due to the malevolence of his enemy. "Every time they tried to give him soup or anything to eat, something would come crawling up in his throat," stated the narrator. After the victim died, witnesses explained that a "turtle come up in his throat and choked him to death." Whether the aftermath of poisoning, or just as often the result of a charmed object hidden somewhere in the proximity of the victim, *Hoodoo* infestation was always accompanied by great physical suffering.[11]

Hoodoo sickness narratives generally conform to a pattern in which the body is conceptualized as a site of disease, organic dysfunction and/or pathology. Even as it is porous within, the body external is also open. Permeable by nature and supernature, the external body, the material self, continually interacts with innumerable invisible forces and powers. Although black Americans generally understood the sources of many common diseases, many also believed that specific disorders could only be caused by spiritual practices such as witchcraft or *Hoodoo*. *Hoodoo* was suspected when an ailment was chronic or unresponsive to treatment; when sickness manifested especially unusual symptoms; or when the illness was dramatic, sudden, and unanticipated. In reviewing the instances of victims and observers who had witnessed bodily intrusions of worms, reptiles, and insects, one is struck by the futility of conventional medical techniques and practitioners in alleviating the suffering of victims. "They's lots of those kind of cases the ord'nary person never hear about," said former bondsman William Adams when questioned. "You hear of the cases the doctors can't understand, nor will they 'spond to treatment." "...the medical doctors aint no good for nothing lak that," asserted Jasper Millegan, an ex-slave whose uncle had been poisoned by a Conjurer's charm, he claimed. "He nebber did get any better and he died." *Hoodoo* afflictions, some blacks believed, could not be corrected by conventional medical means. In general, while healing with herbs, roots and other natural remedies was practiced for common physical ailments, supernatural healing was implemented when other methods were perceived as inadequate. A former bondswoman in Yamacraw, Georgia, for example, argued for the effectiveness of supernatural remedies in especially stubborn cases of sickness. She believed herself to have been "tricked" or *Hoodoo*ed, and ordinary medicine had failed her: "I call a regluh doctuh," she declared, "but he didn seem tuh do no good." She was encouraged by the treatment that the folk healer provided for her. "...three weeks ago I went tuh a root man," she stated hopefully, "He gimme sumpm tuh take an sumpm tuh put in muh bed...I guess I soon be well."

Hoodoo narratives appear to be much more than idiosyncratic descriptions of disease and illness symptoms, for more often than not they provide alternative explanations for undeserved physical suffering. The idea that spiritual causes and effects are at the heart of affliction reflects a personal cosmology. In African American *Hoodoo* practices and beliefs, sickness by infestation was understood to be the outcome of deadly intent and malicious human motivations. The direct products of *Hoodoo* were physical, the means, spiritual. The worldview out of which such ideas emerged was certainly archaic, transcending religious particularity and sectarian beliefs. As part of this traditional vision, the body was envisioned as a battleground in which good and evil forces fought for sovereignty, while pain and infestation exhibited the carnage of a greater war which manifested within the self. In African American understandings of *Hoodoo*, conceptions of illness, disease and other ailments were joined to non-physical sources, such as the acts of an evil, ill-willed enemy, or the destructive powers of witchcraft or sorcery. Illness, misfortune and adversity were never random, but were explicable within the broader scheme of cultural beliefs. Some ailments were believed to be the product of malevolent forces, and some were thought to be sent from God. But many afflictions found their source within the realm of human relations.[12]

It is true that social factors figure prominently in the ways that African Americans experienced and interpreted bodily infestation. In this scheme, illness and disease encompassed social and spiritual processes. Many black folk, for example, indicated that *Hoodoo* illnesses was the result of specific conflicts, tensions, dissension, and stress that engaged community members. Significantly, *Hoodoo* attacks usually manifested in the wake of disputes between individuals. The narratives also suggest that *Hoodoo* poisoning followed the lines of kinship and family relations, striking members of the same household, individuals who had been romantically involved, or persons taking part in a particularly competitive activity or endeavor. The physical and spiritual consequences of *Hoodoo* allegations were eminently social, and in nearly all of the cases I have reviewed, *Hoodoo* illness and affliction ensnared both victims and witnesses in a web of allegations, anxieties, and suspicions of jealousy, anger, and revenge.

The most fascinating and unique element of African American *Hoodoo* tradition is the metaphor of infestation that is articulated and embodied in the narratives. Patterns in the stories of infestation suggest that witnesses sometimes collapsed distinctions in their attempt to give form to the experience of the afflicted. For example, "snakes" were described as "miniature" and sometimes "larval," conceptually displacing the more prominent and imposing serpent, a creature that appears very frequently in American folklore as a symbolic agent of sin and danger. In some cases, the graphic images of worms, maggots, and salamanders seem to be conflated as generalized descriptors, irrespective of the particular characteristics of their different species. It is challenging to identify the origins of these diverse styles and types of infestation since so many *Hoodoo* stories assumed local explanations. In North Carolina, for instance, author Charles Chesnutt believed that stories of animal intrusions originated with members of rural populations, who might have analogized their own suffering as victims of burrowing subcutaneous insects and parasites with the more extravagant claims of Hoodoo infestation. In Mississippi, Newbell Puckett identified a potential source of these ideas in the visual and metonymic associations between common intestinal parasites and tapeworms. Likewise, popular science writers in the nineteenth century noted parallels with respect to patients who endured the pain of bowel infections, such as giardiasis. One comment, from an amateur naturalist at the turn of the century, insisted upon the visual proof of such an infestation, a case that had "reconciled animals of the coldest and most meagre habits to the enjoyment of the warmth and luxuries of the human stomach." Like a witness to a *Hoodoo* consultation, the writer recorded the events "as they actually occurred." "On my arrival, I found that she, (the patient,) had puked up two ground puppies and was labouring under a violent sick stomach, with pain, and syncope:"

> These animals have, since, been shown to me: they are not the ground puppy, (gecko,) as they are vulgarly called; they resemble it very much, but are easily distinguished from it; they belong to the same genus, (lacerta, or lizard,) but are of the species " salamander ;" these animals though oviparous, hatch their eggs in the belly, like the viper and produce about fifty young at a birth. The inference is irresistible, that the patient had, in her frequent draughts of swamp water, swallowed, perhaps thousands of these animals in their nascent, or most diminutive state of existence, and a few only survived the shock; their being, and for which, they were constituted by nature, they should bear this sadden transportation to a situation so opposite in its character, and grow into vigorous maturity, unannoyed

> by the active chemical and mechanical powers to whose operation they were subjected.but it is matter of astonishment, that from the icy element in which they had commenced.[13]

Other observers characterized the presence of entities inside of the body as the outward manifestation of a victim's inward mental and emotional state. A Mississippi police chief told a folklorist in the early twentieth century of an African American woman who had experienced "great agony" because she believed that she had been '*Hoodoo*ed.' "She was screaming and hollering," he said, "[as she] imagined that she had lizards and various other things creeping under her skin and flesh."[14]

While adopting metaphysical categories in order to explain affliction, African-Americans relied upon both conventional and unusual techniques for healing. Since the powers that were believed to have caused creature intrusions were diverse, they could only be controlled by individuals possessing the requisite skills and techniques for curing. The removal of an afflicting supernatural force or power could occur by several means. "I saw a root doctor cut out of a man's leg a lizard and a grasshopper, and then he got well," claimed the renowned root doctor of Columbia County, Georgia, Braziel Robinson. In North Carolina Charles Chesnutt recounted a story of a man who required surgical intervention after he had been poisoned by a lizard. "This lizard, according to the "doctor," would start from the man's shoulder and pass down the side of the body to the leg," Chesnutt wrote. "When it reached the calf of the leg the lizard's head would appear right under the skin. After it had been perceptible for three days the lizard was to be cut out with a razor, or the man would die."[15] For some *Hoodoo* doctors, a modified form of leeching or cupping extraction was the only way to remove offending animal entities. One writer described the process as she witnessed it on the South Carolina Sea Islands in the late nineteenth century:

> There are one or two herb doctors on the island who profess (with the aid of a magical cow horn and certain herbs) to be able to remove the witchcraft out a the person bewitched; and so credulous are the people on this subject that, when the doctor pulls old lizards, etc., out of the horn (previously put there by himself) and asserts that he has drawn them from the sufferer, they not only firmly believe this to be the case, but will describe in extravagant language the feelings experienced during the operation.

A similar account, recorded by a skeptical but bewildered witness in early twentieth century Charleston, SC, describes the process in nearly identical terms. The afflicted patient sent for a Conjure doctor, who immediately took charge. "Get me two plates of salt," he commanded the surprised caregiver. "He pricked the back of [the patient's] neck, drew blood, and ...put his cupping horn to it...He took it off, and dropped out of a young snake and a lizard, dropped them into the salt. They squirmed about – lizards and snakes don't like salt."[16] The dramatic and often elaborate protocols involved with removing infestations from the body demonstrate that treatment was seen to be as complicated a practice as magical poisoning and supernatural affliction. Together, healing and harming formed the most prominent expressions of the expertise of African American *Hoodoo* practitioners.

Meaning

Hoodoo beliefs assigned cultural significance to sickness by mapping the body, both literally and metaphorically, beyond the self. The body, as an infested container of sickness and disease, marked the boundaries between physical disorder and order, spiritual imbalance and wellness. As such, the Hoodooed body mediated the visible and the invisible realms. African American stories of affliction can be thus read as compelling and original interpretations of physical and spiritual dysfunction, much like conventional medical interpretations that posit the invasion of millions of microscopic foreign substances as the source of an illness. In the case of *Hoodoo*, the foreign substances are given explicit form, and the body becomes something of a living menagerie. African-Americans have historically maintained distinctive ideas concerning the meaning of infestation-related sickness, its origin and its cure. These particular notions of illness are expressed with a language of supernatural causality. In their characterization of affliction, blacks paid particular attention to the spiritual dimensions. This does not mean that they dismissed as meaningless physical etiologies in their quest for explanations of disease. It is, rather, that their views of illness were more inclusive of external agencies. In other words, illness was identified by its specific physical symptoms and linked to other sources that were in turn situated in the relational context. Illness, in many cases, was believed to hold social implications.

Nevertheless it is clear that with *Hoodoo* infestations, affliction is much more than the physical symptoms that are so incisively described as bodily states. Affliction represents an attack by an invisible agent, motivated by human intent and action. Social conflict, then, is expressed in the bodily state. The body, in this scheme, reflects a cosmology by which experience (the physical and social realm) and the spiritual world (invisible and visible) gain manifestation. The metaphysical realm enclosed and was enclosed by the body, but paradoxically, it was inclusive of the world in which persons lived and acted. *Hoodoo* affliction was thus located at the nexus of spiritual perception and physical experience. These ideas point to what might be considered dual social and spiritual etiologies of illness in African American life.

Anthropological theory may help to explain the persistence of these ideas in black communities since slavery. Like witchcraft allegations across cultures, experiences of Hoodoo reveal frayed areas in the fabric of social relations. Allegations of Hoodoo affliction become more salient when persons find themselves grappling for power in circumstances that are defined by finite means and limited autonomy. Given the history of American racial oppression, concerns for economic mobility, protection and safety were foremost for African Americans since their sojourn as an enslaved people in the United States, and *Hoodoo* was often allied with achieving or not achieving these goals. For instance, infestation illnesses emerged in relation to accusations that were based in economic rivalry, property disputes, and sexual competition, and these were often directed against persons who were viewed as attaining opportunities at the expense of others. In black communities that perceived themselves to be under siege, greedy and ambitious individuals may have represented a discomforting threat to others – particularly when all were expected to share resources in the struggle for advancement. Ambition itself was not an evil, but striving for individual, selfish gain – as opposed to group uplift and communal progress – may have fostered resentments that led to supernatural affliction and Hoodoo accusations. Hoodoo, then, may have functioned as an expression of social control for African Americans in a world in which personal aspirations vied with collective interests, and where cultural forces and religious attitudes shaped community norms.

The infesting creatures themselves may also provide a clue to the meaning of *Hoodoo* affliction. The usual species of animal invaders described in *Hoodoo* stories included worms, turtles, spiders, scorpions, frogs, ground puppies (salamanders), and snails – all animals associated with mud, water, or slime. Their presence beneath the skin, infiltrating the cavities without restraint, overwhelmingly present in the body but very clearly not of the body, suggests an image of a human container. Porous, without boundaries, the inner person is overrun and corrupted by that which is alien to it. The Conjured body is open to powerful external forces, even as these forces are manifested and enclosed from within. It might be possible, then, to view African American bodies that were afflicted by Hoodoo as symbolizing defilement by an unseen - and particularly dangerous - trespasser. As the anthropologist Mary Douglas has argued, the human body reveals a terrain of "natural symbols," a site of complex social codes and metaphysical commitments. But it is her view of impure and offensive transgressors that is most interesting here. The "unclean animal," she notes, "is that which creeps, crawls or swarms...whether we call it teeming, trailing, creeping, crawling or swarming, it is an indeterminate form of movement..." The religious association between unclean creatures and impurity suggests that African Americans might have classified certain types of illness in analogous ways, viewing infestation as a moral sign. Sickness and disease embodied pollution. In the cosmology of sickness, then, infestation from Hoodoo might have signified social decay and cultural disorder – and a threat to communities and relationships whose boundaries were perceived as dangerously unstable or vulnerable to the outside world.

Hoodoo infestation stories are much more than folklore or fantasy. They offer dramatic historical documentation of embodied pain, and tell us something about the underlying perspectives of many African Americans with respect to notions of spirit and body, sickness and wellbeing. Reading these narratives of the body allows us to see how African Americans may have viewed sickness as a multifaceted experience that could be addressed and interpreted in a variety of ways. In the cosmology of *Hoodoo* sickness, pain was not an abstraction, just as personal suffering was rarely attributed to chance or divine will. Unanticipated bad health had broad meanings and manifestations. Intrusion by worms and worm-like invaders was a sign for more complex and insidious experiences of affliction. *Hoodoo* infestation could be the consequence of the destructive powers of witchcraft, or as often it could be the product of the evil thoughts of a mean-spirited person, or directed through the noxious medium of a Conjurer's charm. *Hoodoo*, I believe, was a shared formulation by which African Americans managed sickness, and by extension, they were able to manage the forces that they believed to have eventuated in the bodily infirmities with which they suffered. In slavery and in freedom, the perpetuation of such culturally-specific ideas of health and sickness ensued largely from the myriad social concerns that they addressed. To take African American *Hoodoo* narratives seriously is to come closer to understanding how healing and health, embodied as knowledge, as spirituality, and as practice - is placed in the service of lived needs and values in every moment and amongst every people.

Notes

1. Anonymous, *Conjure,* Robert Tallant Folklore Collection, 1940
2. Gloria Naylor, *Mama Day* (New York: Vintage Press, 1989), 287–290.
3. There are few studies that treat embodied illness in African American culture, and none that comprehensively treats Hoodoo as an illness system with religious referents. I have been inspired by the writings of Albert Raboteau in this area, especially his "The Afro-American Traditions," in Ronald Numbers and Darryl Amundsen, *Caring and Curing*, (Baltimore: Johns Hopkins University Press, 1986). In another influential text, Sharla Fett examines African American conceptions of health and healing in the antebellum South and treats Hoodoo as a cultural practice that emerged from "a relational vision" of community and health. I would argue, as Fett does, that harming practices also supported this "relational vision" and the social basis of affliction among enslaved blacks. See Sharla Fett, *Working Cures: Healing, Power and Health on Southern Slave Plantations*, (Chapel Hill: University of North Carolina Press, 2002).
4. *Southern Workman* 7, No. 4, April 1878, 31 (*Strange Ways and Sweet Dreams*,239)
5. On Hoodoo narratives as oral histories, see Yvonne Chireau, "Natural and Supernatural: African American Hoodoo Narratives of Sickness and Healing," in Stephanie Mitchem and Emilie Townes, *Faith Health and Healing in African American Life* (Santa Barbara: Praeger, 2008).
6. Georgia Writers Project, *Drums and Shadows: Survival Studies Among the Georgia Coastal Negroes*, (Washington DC: Work Projects Administration, 1940), 13; Newbell Niles Puckett, *Folk Beliefs of the Southern Negroes* (Chapel Hill: University of North Carolina Press, 1926), 252.
7. *Southern Workman* 7, No. 4, April 1878, 31. The variety of animal intruders in Hoodoo narratives is astounding. My research of texts has considered at least twenty varieties of

species, including worms, spiders, scorpions, grasshoppers, undesignated insects, frogs, turtles, slugs, salamanders, snakes, lizards, roaches, snails, guppies, minnows and other undesignated fish, crabs and leeches. For the multiple uses and meanings of creatures in African American Hoodoo traditions, see the collected works of Harry Middleton Hyatt, *Hoodoo-Conjuration-Witchcraft-Rootwork*, Volumes 1–5, (self-published) 1978.

8. Alice Bacon, "Conjuring and Conjure-Doctors," *Southern Workman* 28, 1895, 210.
9. On the rise of Hoodoo as a subject of folklore study, see Ronald Waters, *Strange Ways and Sweet Dreams, Folklore from the Hampton Institute*, 1983.
10. For the history of Black American Hoodoo and Conjure traditions, see Theophus Smith, *Conjuring Culture: Formations of Black America* (Oxford: Oxford University Press, 1994); Yvonne Chireau, *Black Magic: Religion and the African American Conjuring Tradition* (Berkeley: University of California Press, 2000); Jeffrey Anderson, *Conjure in African American Society* (Baton Rouge: Louisiana State University Press, 2005).
11. Octave Thanet, "Folklore in Arkansas," *Journal of American Folklore*, 1892; George Rawick, *The American Slave: an Autobiography*, Vol. 13, 277. On the religious and historical sources of infestation, see Yvonne Chireau, "Body as Menagerie: Sickness and African American Hoodoo Narratives," *Council of Societies for the Study of Religion Bulletin*, February 2007.
12. Charla Fett, *Working Cures: Healing, Power and Health on Southern Slave Plantations*, 2002.
13. Joseph Muse, "Notice of the appearance of Fish and Lizards in extraordinary circumstances," *American Journal of Science* 64, 1829.
14. Puckett, 253–254.
15. Roland Steiner, "Braziel Robinson Possessed of Two Spirits," *Journal of American Folklore* 13, 1900, 228; Chesnutt, "Superstitions and Folklore of the South," in *Mother Wit from the Laughing Barrel* (Jackson: University Press of Mississippi, 1990) 373.
16. Portia Smiley, "Notes and Queries," *Journal of American Folklore* 32 1919, 379. I have documented the cupping technique amongst slave healers in the late 18th century, suggesting a possible African source.

We Are What We *Don't* Eat: Worms, Bacteria, and the Soil Under Us

Todd LeVasseur

This paper will explore the status of worms, soil bacteria, and soil microbes within competing and antagonistic Western approaches to agriculture. It will also illuminate the differences between the two religious worldviews that inform these approaches. One of these worldviews sees worms and bacteria as invaders and pests that have no intrinsic value, being only worthy of extermination; the other sees worms and bacteria on a positive continuum that starts with simply valuing their potential profitability and culminates in seeing them as beings worthy of respect and reverent care as the foundations of soil health and as such are part of our biological and spiritual ken.

Builders of Humus

Earthworms, beneficial soil microbes, and soil bacteria play an important role in systems of organic agriculture. The term "organic" raises a host of issues but here it will refer to the definition given by the United States Department of Agriculture in their USDA Organic guidelines. These defining USDA guidelines are complex, with various standards for both livestock and produce.[1] In order for crops to be certified organic, the following demands must be met: farm land must go through a three year transitional period where banned/non-organic substances are no longer applied to fields; soil fertility must be managed by: cover crops, rotational planting, application of waste materials from animals and crops (i.e. compost), tillage and cultivation regimes, and the application of allowed synthetic and non-synthetic materials; the preferred use of organic seeds; and control of pests, weeds, and diseases by physical, mechanical, and biological management practices. To be considered for organic certification, livestock must be given access to pasture and the outdoors, be fed organic feed, and must not be given antibiotics or growth hormones.

It must be noted that there are those within the organic food community (retailers, consumers, growers, processors) who claim that this definition of organic has been diluted and even co-opted by corporate agribusiness interests seeking to profit on the bourgeoning organic food industry. Others feel that this definition is not strict enough, and that prior to the USDA assuming regulatory control over the definition and guidelines of the label organic, dictated by federal legislative fiat, the term's meaning and the farming methods covered by the word "organic" were purer and stricter.[2] These vocal opponents of the government role in organic agriculture are now looking to move "beyond organic."[3]

Regardless of where U.S. citizens might fall in this debate, most every supporter of organic agriculture agrees with some basic tenets of organic farming practices, most of which are included in the USDA guidelines mentioned above. These practices include companion planting, crop rotation, growing a polyculture of crops/plants, humane treatment of animals, creating habitat on site for various animals (worms and microbes included), and minimal to no use of synthetic and petroleum based pesticides, herbicides, fungicides, and fertilizers.[4] Of course, even within this general consensus, there are at times acrimonious debates between various vested interests in what constitutes organic farming: for example, does a large scale organic farm growing 300 acres of a broccoli monoculture for Wal-Mart count as organic, even if it is USDA certified?

Despite these conundrums, the above practices of organic agriculture both depend on and ideally attain one of its main goals: soil health. To this end, many monographs and studies emphasize the importance of worms and bacteria in creating a robust, healthy organic agriculture.[5] However, most of us, if we have any experience with worms at all, have most likely seen them dangling on the end of a fishing line. Equally, most of our experience with bacteria is clouded by messages from the media, whether news or advertising, that warn us of the myriad harmful side effects of bacteria and dangerous microbes. However, to grow healthy, organic food, according to Sir Albert Howard's 1940 seminal classic *An Agricultural Testament* and multitudes of similar-themed books published since 1940 and devoted to delineating the basics of organic agriculture, both worms and healthy soil bacteria are a godsend.[6]

The common varieties of earthworm and night crawler that we encounter in American soils disembarked with early colonists at Jamestown. These worms were carried over in root balls or dumped on land as part of ballast. Their introduction changed the landscape, especially of Eastern woodlands, because prior to the earthworm, there was nothing that ate the leaves that would fall to the ground every autumn.[7] Besides eating leaves, thus recycling nutrients stored in the leaves back into the soil and providing nutrients to growing flora, earthworms also aerate roots, provide tunnels for rainwater to flow through, and excrete pellets of nutrients throughout the soil. As the Indian organic food activist Vandana Shiva puts it, "The little earthworm working invisibly in the soil is actually a tractor, fertilizer factory, and dam combined."[8]

There are dung earthworms, earthworms that thrive in compost, and still others who live in pastures rather than in gardens. Worms that live higher in the soil have more offspring than worms living lower in the soil—in good soil, one surface dwelling worm can have 1,000 offspring in a year (of course not all of these will survive). These varieties of worms appear in ever increasing numbers of studies that show the role worms have in creating and maintaining healthy, fertile soils, studies which inspire growing numbers of organic farmers to create optimal conditions on their farms for worm habitat. In fact, vermiculture—the practice of growing and breeding worms either for the worms themselves or their castings—is a rapidly growing business in the organic agriculture industry.

Besides worms, there are various species of bacteria and microbes in the soil that also play an equally important role in organic agriculture. Although mainstream advertising and our cleanliness-obsessed culture disparagingly label bacteria and microbes in general as parasitic and dangerous to human health (and certainly some are, especially if manure based compost--a central ingredient to almost all forms of organic agriculture--is not properly broken down), soil bacteria and microbes nonetheless play a pivotal role in creating and maintaining soil health on organic farms. For organic farmers, soil health is equated to the presence, abundance, and vitality of soil microbes that result from the above mentioned practices of organic agriculture (especially the application of well made compost). Many organic farmers and consumers of organic agricultural products feel that the healthier the soil is, in terms of nutrients available to plants, then the healthier the plants are that are grown in such soils. These farmers and consumers maintain that the humans who eat the plants will be healthier, too. The same rule applies to organic animal husbandry—the healthier the animal (raised on a diet of organic food, allowed to "free-range," and not given antibiotics), then the healthier their meat, milk, and other products will be for human consumers.

Soil bacteria and microbes are instrumental in helping agricultural crops feed upon nutrients found in the soil. Well-made compost when applied to organic fields becomes the mineral and nutrient dense soil humus from which crops feed. Worms multiply and course through this nutrient dense humus, helping to aerate the soil and add to soil fertility by releasing their castings, thus improving the nutrient level and humus content of the soil. Microbes and bacteria help this crop feeding process as well, from rhizomic activity around plant roots to the ongoing decomposition of the deceased bodies of bacteria and microbes: these decomposing cellular bodies become food for plants and their decaying biomass helps to further build soil humus. Organic advocates claim that this improved soil humus leads to pH-balanced soil, greater retention of water, and that it provides plants with habitat for healthy, vigorous growth that leaves them less susceptible to blights and pests. Most organic farmers will maintain that none of the above is possible for their crops and the various life forms that create on-farm habitat without worms, bacteria, and microbes.

Given this important role bacteria and microbes (and worms) play in helping to create the literal foundations for life, the famed Harvard biologist E.O. Wilson points out that "Bacteria and bacteria-like microbes called *archaea* are the dark matter of Earth's living universe."[9] He makes this claim because in one gram of fertile soil there exists ten billion bacteria and in one ton of soil there is an estimated four million species of bacteria. These vast amounts of bacteria and microbes literally help create and constitute the soils under our feet and in agricultural fields. Therefore, creating and maintaining this diversity of bacterial life within pastures and crop fields is seen as one of the major goals of organic agriculture.

Proponents of organic agricultural methods attempt to realize this goal through the rituals of making and spreading compost, the creation and casting of biodynamic preps in biodynamic agriculture as spelled out by Rudolph Steiner, the breeding of worms, and the creation of habitat (especially compost piles) for earthworms. All of these activities have as their goal the maintenance and creation of bacterial and microbial life of the soil.[10] If we were to reduce this organic ideal to an equation, it would read: healthy levels of bacteria and worms = healthy soils = healthy crops = healthy humans. Organic farming advocates further claim that this equation contributes to on-farm biodiversity and healthy ecosystems.

The Grass is NOT Always Greener

In contrast to the organic equation (or formula) exists a competing farming worldview. This conventional farming worldview sees worms and soil microbes as enemies that need eradicating, especially if these organisms are deemed a threat to crop yields and farm profits. Supporting this worldview and approach to worms and soil microbes, historically and still in great measure today, are land grant universities, chemical agribusiness corporations, and the US Government, especially through its Farm Bill subsidies.[11] In this worldview, farming is in large part about increasing yields of cash crops for markets and corporate profit and is heavily dependent on machines (large tractors, combines, etc.), hybrid and genetically engineered seeds, and synthetic-chemical intense farming methods.[12]

It is important to emphasize that the distinction between these two forms of farming, and the soon to be explored, religious worldviews behind them, exists for partly heuristic reasons. These categories are not "hard and fast" distinctions, nor do they completely exhaust the range of values practitioners self-report as motivations for farming organically or conventionally.[13] However, such distinctions are necessary for systematic analysis of the values and beliefs that motivate these contrasting forms of farming, in large part because these forms of farming and the worldviews behind them have different material repercussions in the environment. In the case of conventional farming, these include algae and toxic bloom "dead zones" in the Gulf of Mexico from nitrogen-based fertilizer run-off draining out of the Mississippi River; Mad Cow disease, which results from cows in Concentrated Animal Feedlot Operations (CAFOs) being fed ground up body parts of dead sheep and other cows as they are fattened up for slaughter and market; and the poisoning of farm workers who apply agricultural chemicals to the fields of conventionally farmed produce.[14]

The chemicals used in conventional farming regimes, which critics say pollute the environment and the bodies of carbon-based life forms, are a result, both technological and ideological, of twentieth-century violence. Professor of technology, culture, and communication Edmund Russell has shown how the ideology of waging war on pests in agricultural fields co-evolved with ideologies of waging war on our human enemies in the two World Wars.[15] In effect, he describes how the processes of using synthetic chemicals to bomb human beings during war and in agricultural fields when farming both share the same technologies and chemical ingredients.

This ideology of chemical-backed farming technologies and practices was greatly aided by the revolutionary hypothesis of the German chemist Justus von Liebig. In the late 1800s, von Liebig claimed that plants were almost wholly nourished by nitrogen, phosphorous, and potassium carbonate; furthermore, during this period of his scholarship and research, he maintained that plants were not significantly nourished by any of the other manifold minerals and microbes that constitute the soil humus in which plants are found. Thus, von Liebig almost single handedly overturned the traditional, low-input farming paradigm and practices that were extant prior to that time and opened the door for today's chemical-based conventional farming. Ten years after releasing his initial studies and revolutionary view, von Liebig realized that increased soil fertility actually results from applying organic materials via compost and that the minerals and microbes in the soil are what feed plants, but by then it was too late--profits were being made and the belief about nitrogen, phosphorous, and potassium carbonate being central to plant growth assumed a life of its own.

This "life of its own" occurred when poison gasses left over after WW I were profitably applied to insects on and in the soils of fields in both Europe and the United States, thus beginning the multi-billion dollar petroleum and synthetic-chemical based insecticide, herbicide, and fungicide industries that exist today. Most of the companies that produce these chemical products—Du Pont, Dow, and Monsanto—began by making liquid ammonia that was used in WW II as weapons and explosives. DDT was also created in this era, at first to keep fleas and lice off of GIs, and later to keep insects off of crops.[16] Despite intense lobbying and incomplete and questionable testing methods on behalf of the advocates for these products and technologies, studies empirically show that these chemicals kill life—from microbial to worm to plant to human.[17] Overall, U.S. farmers apply twenty-one million tons of nitrogen, phosphate, and potash a year on the country's 907 million acres of cropland. This tonnage is a combination of 50,000 pesticide formulas and over 700 different chemicals, all geared towards eradicating and controlling pests and blights and boosting yields.[18]

Whither Religion? In the Soil or in the Chemical?

I posit and argue that the organic and conventional forms of farming outlined above, both of which have impacted views about and habitats of worms and microbial bacteria, are part and parcel of two competing religious worldviews. One of these religious worldviews sees the earth as a living system of interdependent parts. Certain followers of and adherents to this worldview even go so far as to say the earth itself and the myriad life forms residing thereon are sacred and worthy of reverent care. Inherent to this worldview is the perception that nature is a teacher and that natural processes and organisms and their habitats are to be respected and cared for. The other religious worldview sees worms and insects as pests needing to be controlled and ideally eradicated and the earth and its biotic life as dead, inert matter, there to be manipulated, killed, and processed in the name of profit. I will now articulate this latter religious worldview before exploring the former.[19]

The religious worldview that motivates the practices of conventional farming sees the earth and its organisms and processes as atomistic background material to be manipulated for profit.[20] This worldview subscribes to and follows what religious studies scholar David Loy terms "The Religion of the Market" (1997).[21] Loy posits that by adopting a functionalist approach to religion, it is possible to conclude that consumerism has become the most powerful value-system of the world today.[22] Loy explains how economics and the theorized functioning of market mechanisms have long since left the realm of being a discipline and profession and today have assumed the air of a religion and function as one. Significantly, for Loy, this religion has as its god the Market (which he writes with a capital "M" to recognize its divine stature). Loy claims that the religion of the Market has become the world's most widespread and dominant religion, offering a secular salvation based on production, consumption, and the accumulation of profit. Furthermore, its eschatological worldview is of a future capitalist utopia with individual bank accounts magically filled by retirement age--Loy maintains that this end-times belief is aggressively pitched around the world. Loy ends his treatise on this religion by claiming it is to our common detriment (academically, politically, socially, and ecologically) if we only view the Market in secular, rather than religious, terms.

Especially important for the theme of this paper, Loy writes that, "The degradation of the earth and the degradation of our own societies must both be seen as results of the same market process of commodification—which continues to rationalize its operation as natural and inevitable."[23] Here we see that The Market, operating through its "priestly class" of corporations, advertisers, and government supporters (Loy claims that the U.S. Government is the most powerful proponent of apotheosized Market capitalism) advances under a worldview of quarterly profits earned via trading commodities on international markets.[24] This view of the market--where the earth becomes lifeless commodities to be traded and sold--trickles down to farm policy decisions and farm management practices that have resulted in an "unsettling of America," to use Wendell Berry's seminal phrase, and in a symbiotic increase in conventional farming practices. These farming practices are heavily dependent on chemicals, crop patterns, tillage methods, and other farming technologies germane to conventional farming, all of which critics claim are inherently harmful to worms, soil bacteria, and microbes in the soil. The goal of these farming practices is to boost yields of monoculture crops that can be sold on national and international commodity markets. The soil thus becomes a medium that is injected with deadly chemicals, fungicides, pesticides, and fertilizers so that photosynthesizing crops can be grown, processed, and traded with the goal being the accumulation of profit, as this is what the Market demands of its followers. As Nixon's Secretary of Agriculture Earl "Rusty" Butz famously said to farmers: "get big or get out."[25] In other words, play by the rules of the Market, or lose your farm.

In this religious worldview of the Market, the act of farming and the inhabitants of farming soils are strictly viewed in utilitarian terms as a way to make profit. The actual care for and stewardship of farmland and inhabitants of farm soils--such as worms—only registers in this religious worldview if it helps to make money. Ironically, as seen with the exponential growth of organic foods, there is money to be made in organic farming. One problem with this is that chemical companies and large agri-business corporations, such as Monsanto, will be unable to sell their products in the organic paradigm, creating a clear disincentive to support worm and bacteria friendly farming practices. As Mark Winston explains:

> Pest management is a business. It has science behind it, sometimes fascinating science, but in the end the bottom line determines whether a piece of interesting

> biological research is relegated to the textbooks or becomes a commercially useful system to control a pest. The farmers, exterminators, extension agents, and agricultural product distributors who make decisions concerning pest control are not swayed by elegant science, clever techniques, or trendy new ideas. Rather, decisions are made by the simplest and most pragmatic criteria: which product or method does the best job of controlling a pest with the minimal cost. Pest management today is still pesticide-heavy because chemical pesticides are the most efficient and direct way of meeting the joint standards of high efficacy at low cost. There have been innumerable scientific advances in developing alternative, biologically based, environmentally friendly solutions to pest management in our century, but none of them has even approached the commercial success of pesticides.[26]

I want to offer two quick addendums to this religion of the Market-inspired approach to farming. The first involves religion scholar Richard Foltz's recent ten-year revisit of David Loy's original article about the religion of the Market. Foltz argues that Loy's article, premises, and conclusions about the religion of the Market are accurate and have strong explanatory power and that, at the least, "the Religion of the Market is surely worthy of broader attention from religion scholars than it has been getting up to now."[27] The second is the observation made by Vandana Shiva of the worldview created by and thus inherent to the religion of the Market. She describes this worldview as being a "monoculture of the mind," where all diversity (in terms of thought, culture, and economic production) is whittled away and colonized so that all that is left are cultural and ecological monocultures, able to be bought, packaged, traded, and sold in the global economy (1993). It can be argued that this monoculture of the mind manifests itself in monocultures of conventional farming landscapes and the practices that shape them: a landscape where the worm and the desire for healthy soils full of bacteria have no room or sway.

I now revisit the worm and bacteria friendly organic farming methods outlined earlier and explore a variety of religious worldviews that underlay these practices. One tributary of this organic farming worldview is a Christian based "Theology of Sustainable Agriculture" argued for by the agricultural engineer Steven Hall.[28] Combining his lay-Protestant reading of scripture with his Ph.D. in agriculture engineering, Hall argues for implementing a Sabbath rest so soils can recover vital nutrients and bacteria and worms can replenish themselves. He also champions a living wage be paid to farmers who practice sound farming practices; a restorative agriculture that heals the soils and organisms damaged by conventional farming practices; and a hoped for cultural redemption of both ourselves and the soils around us through sound farming practices. This is one example of a Christian based organic farming worldview that is worm and bacteria friendly.[29]

Other religious worldviews motivate some practitioners of organic, worm-friendly agriculture.[30] For example, religious historian Rebecca Kneale Gould undertook ethnographic research with a group of New England homesteaders who purposively turned their back on the consumption-based lifestyle of the Market (2005). In the process of trying to "escape" the religion of the Market (my terminology, not that used by her research subjects), some of these homesteaders integrated organic farming into their homesteading practices. Gould found that these homesteaders were motivated by a "psychospiritual longing: to find a form of spiritual life outside formal religious institutions and to practice a way of living that is in step with the natural world rather than against it" (ibid: xx). This quote shows that, at least for some homesteaders and others who practice organic agriculture, a life lived close to nature and at odds with dominant urban consumer lifeways can be *explicitly religious*, with nature, nature's processes, and respect and reverence for nature's creatures and organisms at the center of that religion. In support of this claim, Gould found that reverence for nature tops the list of attributes that constitute the religious worldviews of the homesteaders she interviewed.[31]

Continuing with this exploration of possible religious worldviews and motivations behind the practice of organic worm and bacteria friendly agriculture, I offer religion scholar Bron Taylor's recent scholarship that posits the existence of an emerging "Dark Green Religion" (2010).[32] He claims that this nascent religion has its roots in various religious environmentalisms and that it is significantly impacted by insights from ecology and evolutionary biology.[33] In dark green religion reverent care for and appreciation of the biodiversity of life in all its shapes and sizes is a central tenet. Of the many characteristics that Taylor claims constitute dark green religion and other nature-as-sacred religions, namely, "religions that consider nature itself to be inherently sacred," is "the biocentric turn" (1994) that these forms of religion have adopted. For Taylor, these emerging nature religions represent a religious "turn toward values professing that nature has intrinsic or inherent value," (2010) as compared to the religion of the Market, where nature has no value except what the Market dictates.

This biocentric turn Taylor describes can be seen in the kinship ethics that many organic farmers embody in their attempt to practice sustainable and earth friendly agriculture. Such kinship ethics can and may even include viewing other life forms in nature and in the soil as spiritual kin; as fellow travelers in the unfolding process of evolution on this planet; and even as spiritual teachers and powerful beings in and of themselves, especially for those practitioners of dark green religion who Taylor claims hold an animistic worldview.

A final example of religious motivations and approaches to worm and bacteria friendly farming comes from the organic farmer and agrarian essayist Wendell Berry. Berry writes that, "Because the soil is alive, various, intricate, and because its processes yield more readily to imitation than to analysis, more readily to care than to coercion…agriculture…is a practical art. But it is also a practical religion, a practice of religion, a rite."[34] In Berry's agrarian understanding, the soil is alive, as are the organisms and worms in the soil. In this worldview, it is proper to care for these life forms—in fact, to do so, according to Berry, is both a religious duty and a moral obligation.

Beyond the above discussion of conventional and organic, gray areas exist within the farming and religious motivations behind these different practices. To be sure, there are conventional farmers who are trapped in a cycle of debt who might no longer wish to commit farming acts that kill soil microbes and worms. Just as surely, there are organic farmers who view worms and compost in strictly utilitarian terms and who are only in organic farming for money, or for personal health issues, and therefore might have no biocentric kinship ethic motivating their farming. In fact, one of the main findings of Gould's book is the ambiguity and ambivalence homesteaders have toward nature despite psycho-spiritual longings that for some includes forming a spiritual bond with nature.

Ambiguity aside, the worldview farmers bring to their engagement with nature impacts how they interact with it and how they farm. If they see the planet as inert matter to be traded for profit, then worms, soil bacteria, and soil microbes will suffer; if they see a living planet that is to be spiritually worshipped and whose biological processes are to be subtly directed for health, wealth, and harmony, then worms, soil bacteria, and soil microbes will thrive.

Conclusion: Harvesting the Fruits of Our Labor (and the Labor of Worms and Bacteria)

The explorations of this paper are emblematic of a larger process of America's history of creating contested sacred space. This process is expertly outlined by scholars Edward Linenthal and David Chidester, who write that "sacred space has been ritualized, reinterpreted, and contested all over America."[35] It should be of no surprise that this process is occurring on the farmland of our country, too. In the end, this means that worms and bacteria not only help create and maintain the health of our soils, and by default our bodies, but they are unwitting victims in a religious battle being waged by competing religious worldviews on contested ground throughout our country. On one side is Loy's Religion of the Market, which sees the earth and its life forms as a monoculture of products to be sold at the altar of Mammon; on the other side sits various "nature as sacred" strands of Christianity, Judaism, religious environmentalisms, and emerging dark green religions that see the earth and its life forms as being holy, "good," interconnected, and worthy of reverent care.

Postscript: Fornication and Worm Reverence

A few years ago, organic farmer Michael Ableman traveled the country. His journey took him from his own farm on Salt Spring Island in British Columbia to visit U.S. farmers who are practicing the sort of worm and bacteria friendly organic and biodynamic farming practices outlined in this paper. He visited farmers in all parts of the States, including those living in dense urban areas. His trip resulted in his book, called *Fields of Plenty: A Farmer's Journey in Search of Real Food and the People Who Grow It* (2005). One of the farms he visited was Knoll Farms, one of the earliest California Certified Organic Farms. The following is a summary quote he gave regarding his time spent with Rick and Kristie Knoll, the founders, owners, and farmers of Knoll Farms. Notice in this quote both Rick's view on worms, and also Ableman's description of alternative, earth friendly farming methods and how his description borders on a religious testament: "When Rick proudly tells me that there are 'night crawlers fucking in the orchard,' I understand. To know that numerous creatures have settled in and are comfortable enough to reproduce themselves provides as much a sense of accomplishment and satisfaction as high yields and tasty crops. Once again, I'm reminded that farmers who are outlaws by conventional terms are actually lured by a sense of being part of something much bigger than themselves."[36] Whether this something "much bigger" that Ableman praises is a concept of Christian stewardship, of back to the earth religiosity, of dark green religion, or some other form of nature centered worldview, it nonetheless impacts the way worms and bacteria are treated in the soils of agricultural fields around the country and world.

Notes

1. http://www.ams.usda.gov/AMSv1.0/NOP This is the gateway link to the USDA National Organic Program. Accessed on September 12th, 2009. The following web-site has a link to synthetic and non-synthetic substances that can and cannot be used on certified organic agricultural lands and in certified organic products: http://www.ams.usda.gov/AMSv1.0/ams.fetchTemplateData.do?template=TemplateN&navID=NationalListLinkNOPNationalOrganicProgramHome&rightNav1=NationalListLinkNOPNationalOrganicProgramHome&topNav=&leftNav=&page=NOPNationalList&resultType=&acct=nopgeninfo Accessed on September 12th, 2009. This third web-site has an introductory list of certified standard practices: http://www.ams.usda.gov/AMSv1.0/getfile?DocName= STELDEV3004445 Accessed on September 12th, 2009.
2. The legal fiat came with the passing of The Organic Foods Production Act of 1990, which required the Secretary of Agriculture to create and establish a National List of Allowed and Prohibited Substances. Prior to this Act, organic certification was conferred on a state-by-state basis and was administered by various state and regional-level governing bodies; for example, the California Certified Organic Farmers (CCOF) handled certification for California and other regional farmers. The USDA has formed partnerships with these state-level certifiers so that the state-level certifiers help certify and monitor organic farms for the USDA.
3. See, for example, the long-running monthly magazine *ACRES USA: The Voice of Eco-Agriculture* (www.acresusa.com) and its criticisms of the USDA and of corporate interests in today's organic markets, and the work of farmer Joel Salatin, *Everything I Want to Do is Illegal: War Stories From the Local Food Front* (Swoope, VA: Polyface, Inc., 2007).
4. In contradistinction, farmers who do not follow organic practices but rather who rely almost solely on petroleum and synthetic based artificial chemicals, who plant hybrid seeds, who use genetic engineered seeds and other genetic engineering technologies, and who plant monocultures of hundreds to multiple thousands of acres are called "conventional" farmers in modern agricultural parlance. These practices form the core of what is called the "Green Revolution" in farming, where production has shifted from

small scale family farms using traditional farming methods (now in part called "organic" as a response to the Green Revolution, although organic farming methods have developed in sophistication and efficacy of obtaining increased yields compared to 100 years ago) to large scale and tractor, irrigation, and chemically based monocultures of high-yielding hybrids. For an in depth history of this shift from traditional to conventional farming in the U.S., see P. Conkin, *A Revolution Down on the Farm: The Transformation of American Agriculture Since 1929* (Lexington, KY: The University Press of Kentucky, 2009). For a critique of what this shift has done to agrarian farming communities in the U.S., and the role corporations and the U.S. Government have played in this shift, see Wendell Berry, *The Unsettling of America: Culture and Agriculture* (NY: Avon Books, 1977).

5. D. Ernst, *The Farmer's Earthworm Handbook: Managing Your Underground Money-Makers* (Brookfield, WI: Lessiter Publications) 1995; J. Vandermeer, "The Ecological Basis of Alternative Agriculture." *Annual Review of Ecology and Systematics.* Vol 26, 1995, 201-224. There are even books that explain how to handle worms and parasites that impact the health of livestock and to protect livestock from these parasitic organisms in such a way that organic certification can be maintained. See P. Stockdale, *Living With Worms in Organic Sheep Production* (Canadian Organic Growers, 2009).
6. Bradley, Fern et al., eds. *Rodale's Ultimate Encyclopedia of Organic Gardening: The Indispensable Green Resource for Every Gardener* (Emmaus, PA: Rodale Books, 2009); A. Howard, *An Agricultural Testament.* (NY: Oxford University Press, 1940); A. Howard, *The Soil and Health: A Study of Organic Agriculture (Culture of the Land)* (Lexington, KY: The University Press of Kentucky, 2007); C. Walters, *Eco-Farm, An Acres U.S.A. Primer: The Definitive Guide to Managing Farm and Ranch Soil Fertility, Crops, Fertilizers, Weeds and Insects While Avoiding Dangerous Chemicals* (Austin, TX: Acres U.S.A, Inc., 2003).
7. See C. Mann, "America, Found and Lost." *National Geographic.* May 2007, vol. 211, no. 5, esp. 35-37. There have historically been indigenous species of worms native to the North American continent but most of these species became extinct after the last ice age.
8. V. Shiva, *Monocultures of the Mind: Perspectives on Biodiversity and Biotechnology* (Atlantic Highlands, New Jersey: Zed Books, 1993); V. Shiva, *Stolen Harvest: the Hijacking of the Global Food Supply* (Cambridge, MA: South End Press, 2000).
9. E. Wilson, *The Creation: An Appeal to Save Life on Earth* (New York: W.W. Norton & Company, 2006).
10. Biodynamic agriculture is seen by some as being a potential "beyond organic 'savior'." Although still on the margins in the United States, biodynamic agriculture is a large component of organic agriculture regimes in Europe and Australia. This type of agriculture also does not use synthetic chemicals, sees the farm as a "whole organism," applies composts, plants by stars and other celestial bodies, and applies "homeopathic" tinctures to crops and the soil. A. Podolinksy, *Biodynamic Agriculture Introductory Lectures: Volume 1* (St. Leonards, Australia: Gavemer Publishing, 1996); R. Steiner, *Agriculture (An Introductory Reader)* (Forest Row, UK, 2003). The chemist and biodynamic farmer Hugh Lovel claims that these tinctures rapidly and dramatically boost the population of soil microbes and bacteria. H. Lovel, *A Biodynamic Farm* (Austin, TX: Acres U.S.A., Inc., 2000)

11. For example, see Fahnestock (2005), who shares data about a joint USDA-Monsanto-Purdue University study about combating soybean rust with fungicides and pesticides. A. Fahnestock, "Soybean Rust Strategy: R2D2." *Croplife*. 168 (2005),10.
12. Critics of large-scale conventional farming claim that studies show that large machines destroy the soil and thus habitat for worms and bacteria. The weight of the tractors and combines used on today's farms packs the soil so that the soil becomes rock hard and thus nearly impossible for worms to be able to crawl through. Of course, many organic farmers, and especially large corporate organic farms, use tractors. This is another murky area, as most farmers on either "side" maintain that market demands dictate the use of tractors. E. Faulkner, *Plowman's Folly* (NY: Grosset and Dunlap, 1943).
13. Sociologists Beus and Dunlap found six areas of difference between what they call the competing paradigms of conventional and alternative agriculture: dependence vs. independence; domination of nature vs. harmony with nature [of significant import for this paper]; centralization vs. decentralization; exploitation vs. restraint [see below discussion about religious motivations behind these two farming paradigms]; specialization vs. diversity; and competition vs. community [see again below discussion about religious motivations]. See Beus, C. and R. Dunlap. "Conventional versus Alternative Agriculture: The Paradigmatic Roots of the Debate" in *Rural Sociology*, Winter 1990, 55:4, 590-616.
14. Michael Pollan's bestseller *The Omnivore's Dilemma* (2006) provides a well-documented account of these material results of conventional farming. He also offers an astute analysis of the US Farm Bill and how the subsidies contained therein impact the soils and waters of our nation and hemisphere. Again, to show that there are murky areas, some claim that large-scale organic cattle operations, especially in today's sizeable organic market, are inimical to soil and environmental health, despite the traditional ideals of organic agriculture and husbandry as advocated by pioneers like Howard, Walters, and Rodale. For the poisoning of farm workers, see Stephanie Hicks, (http://www.colby.edu/personal/t/thtieten/ag-pest.htm) who shares that "Mevinphos, an organophosphate insecticide used to control aphids, mites, grasshoppers, cutworms, leafhopper caterpillars and other insects, is applied to a variety of vegetables, such as head lettuce, leaf lettuce, cauliflower, broccoli, and celery" and that the application of this chemical has been linked to illness and death in farm workers.
15. Russell, Edmund P. "Speaking of Annihilation': Mobilizing for War Against Human and Insect Enemies, 1914-1945." *The Journal of American History*, Vol. 82, No. 4 (Mar., 1996), 1505-1529.
16. P. Tompkins and C. Bird, *Secrets of the Soil: New Solutions for Restoring our Planet* (Anchorage, AK: Earthpulse Press, 1998). especially their "Introduction," pgs. xi-xviii. These early chemical and weapons companies are still the major suppliers of chemicals and genetic engineering technologies used by today's domestic and international conventional farmers. Most of these companies are what are now called Transnational Corporations, with subsidiaries in multiple countries and investments in various aspects of national and global agricultural regimes. Lastly, Rachel Carson's *Silent Spring* helped educate the public about the harmful side effects of DDT. While outlawed in the U.S., DDT is still manufactured by Monsanto and is sprayed on agricultural fields throughout Latin America.
17. For a sustained critique of conventional farming, the chemicals used in it, and the impacts of these synthetic chemicals on ecosystem and human health, see A. Kimbrell,

Fatal Harvest: The Tragedy of Industrial Agriculture (Washington, D.C.: Foundation for Deep Ecology/Island Press, 2002.

18. Some scientists dedicate their research to mediating the effects of these chemicals in the fields of landscape ecotoxicology and environmental toxicology. Also, Journals such as *Applied Soil Ecology* and *Soil Biology and Chemistry* do publish studies that recognize the important roles various soil bacteria and worms play in agriculture and advocate for Integrated Pest Management (IPM) best-practices that attempt to minimize the use of chemicals in farming. For a textbook based example of chemical-backed farming, see David Alford, ed. *Pest and Disease Management Handbook* (London: Blackwell Science, 2000), 1, who maintains that "Arable farmers…need to make efficient use of variable inputs such as insecticides, fungicides, fertilizers, seeds and energy." While mentioning public concern over such chemicals, and that there are forms of biological pest control available to farmers, the edited volume nonetheless delineates chemical control regimes to combat pests and plant diseases. Lastly, an example of a conventional farming text reads as follows: "Nematode management [nematodes are an extremely diverse species of roundworms] must be viewed as a preplant consideration because once root infection occurs and plant damage becomes visible there are very few nematode management options available to help resolve the nematode problem and avoid potentially significant crop losses. In this regard, pest and crop monitoring activities are very important considerations for early detection of pending problems. Once the discovery is made that nematodes have colonized plant roots and stunted crop growth, the question is whether it is possible to effectively reduce nematode population levels and restore crop yield potential. At present the only post plant nematicide which can be used in some crops to help resolve an established nematode problem is Vydate (Oxamyl). Vydate is not considered a true nematicide, but rather a nematostat. Nematostats, rather than kill nematodes, induce a narcotic effect which paralyzes the nematode and prevents it from feeding, movement, mating, and other normal activities. The narcotic effect is only as persistent as adequate Vydate concentrations are maintained within soil and roots. Following nematicide application, irrigation and rainfall can dilute and leach toxic concentration from the nematode environment, thereby restoring the nematodes ability to conduct normal bodily functions. As a result, repeated and sequential Vydate applications to soil are required to maintain toxic (narcotic) concentrations" (J. W. Noling, "Nematode Management in Commercial Vegetable Production." http://edis.ifas.ufl.edu/ng004). Organic farmers also have farm management practices geared towards controlling blights and pests, but these practices are holistic and have as their goal healthy soils. Most organic farmers, especially those working at smaller scales, accept that up to twenty percent of what they plant will be lost to insects, birds, deer, and disease and/or will not be of good enough quality to sell to retail outlets. Rather than resorting to chemicals to protect their crops, they plant an excess of crops at the beginning of the season, understanding that some will be "lost." Organic farmers also practice companion planting to help protect against blights and pests.
19. As with the discussion about farming methods, I present these religious worldviews in such a clear-cut way in part as a heuristic tool, but also to show that there are actually significantly different beliefs in these two religions and that these beliefs have a material impact on the subject matter of this paper—worms and soil bacteria and microbes.
20. For an in-depth analysis of how Western science from the enlightenment onwards came to view the world in this way, see C. Merchant, *The Death of Nature: Women, Ecology*

and the Scientific Revolution. (San Francisco, CA: HarperSanFrancisco, 1980 [new preface: 1990]). This scientific view is in large part what the religion of the Market uses as its own view of nature. Merchant goes so far as to claim that "the death of nature" (the provocative title of her book) is "the new value system of the modern world" (ibid: 217).

21. Loy, David R. "The Religion of the Market." *Journal of the American Academy of Religion*, Vol. 65, No. 2. (Summer, 1997), 275-290. For similar arguments and claims about capitalism/the global market functioning as a religious system, see H. Cox, "The Market as God: living in the new dispensation." *The Atlantic Monthly* March 1999, 18-23 and "Mammon and the Culture of the Market: A Socio-Theological Critique" in eds. Madsen, Richard, William Sullivan, Ann Swidler, and Steven Tipton *Meaning and Modernity: Religion, Polity, and Self* (Berkeley, CA: University of California Press, 2002), 124-135. For a theological perspective, see C. Deutschmann, "Capitalism as a Religion? An Unorthodox Analysis of Entrepreneurship." *European Journal of Social Theory* 4(4) 2001, 387-403. For a European social theory perspective, see Hopkins (2001) and John Galbreith (who claimed in 1958 that Americans worshipped the "Great God Growth" [cited in Guha, 2006: 222]) "The Religion of Globaliztion" in ed. Hopkins, Lorentzen, Mendieta, and Batstone *Religions/Globalizations: Theories and Cases* (Duke University Press) 2001, 7-34. For a political science perspective, see Kaplan, Jeffrey. "The Gospel of Consumption and the Better Future we Left Behind." *Orion Magazine* May/June 2008. For a Global South perspective, see D. Pena, ed. *Chicano Culture, Ecology, Politics: Subversive Kin* (Tucson, AZ: The University of Arizona Press, 1998), especially when he talks about the "sacred claims of private property" that are central to the functioning of the Market (252). Religion scholar Bron Taylor has cautioned against calling neoliberal market capitalism a religion (personal discussions). "Earth First!'s Religious Radicalism" in Chapple, Christopher Key, ed. *Ecological Prospects: Scientific, Religious, and Aesthetic Perspectives* (Albany, NY: SUNY Press, 1994), 185-209 and *Dark Green Religion: Nature Spirituality and the Planetary Future* (Berkeley, CA: University of California Press, 2009). His caution is founded in part on asking the question, "Would a person voluntarily self-describe as a practitioner of the religion of the Market?" While I respect his perspective on this issue, I feel that Loy's argument from a functional perspective is strong and sound; and that Foltz's ten-year review and his conclusion (see below) add weight to Loy's original claim about the religion of the Market. Also, Gordon Gecko's "Greed is Good" diatribe aside, it is culturally taboo in many social and survey settings to say, "I worship the Market and money." Nonetheless, given the discourse around economics and the amount of press and energy devoted to market mechanisms, I feel confident with calling such a force a religion, at the least in a functional sense, as Loy does. As Chidester and Linenthal demonstrate (1995—see below), cultural conceptions of religion and of what counts as sacred are contested and not necessarily always clear, and it is therefore possible for humans to operate under multiple and at times competing religious belief systems and worldviews. Given the context, one belief system might assume dominance while another becomes latent. Furthermore, I do not subscribe to the position that humans are always cognizant of what their beliefs are and always act on these based on their own choice and volition. Rather, I agree with Loy's conclusion that it is to our detriment (and that of the environment) if we view the Market in strictly secular terms.
22. Loy, "Religion," 257.
23. Loy, "Religion," 284.

24. Loy, "Religion," 282.
25. Cited in Pollan (2006), 52.
26. M. Winston, *Nature Wars: People vs. Pests* (Cambridge, MA: Harvard University Press, 1997), 100.
27. R. Foltz, Richard. "The Religion of the Market: Reflections on a Decade of Discussion." *Worldviews* 11 (2007), 135-154 (152).
28. S. Hall, "Toward a Theology of Sustainable Agriculture." *Perspectives on Science and Christian Faith.* 54: 2 (June 2002), 1-5.
29. For another, heavily Christian influenced organic farming perspective, see the work of Cornell agronomist G. Fick, *Food, Farming, and Faith* (Albany, NY: SUNY Press, 2008). For an agrarian, sustainable agriculture-friendly reading of the Hebrew Bible, see the work of Professor of Bible and Practical Theology Ellen Davis, *Scripture, Culture, and Agriculture: An Agrarian Reading of the Bible*. (NY: Cambridge University Press, 2009). Lastly, religious studies scholar Sarah McFarland Taylor offers case study examples of Catholic women religious and their sustainable lifestyles and worm-friendly farming practices (including some communities that practice biodynamic agriculture): *Green Sisters: A Spiritual Ecology* (Cambridge, MA: Harvard University Press, 2007).
30. Future scholarship on the motivations cited by those who practice organic farming would do well to investigate religious beliefs and worldviews. Two questions worth exploring are whether organic farmers see organic agriculture itself as a religious practice, and if the goals and ideals of organic farming function on their own as a religion for some organic farmers.
31. R. K. Gould, *At Home In Nature: Modern Homesteading and Spiritual Practice in America* (Berkeley, CA: University of California Press, 2005), 23.
32. Author accessed Taylor's unpublished manuscript at the time of this article's acceptance for this book. All quotes are taken from this manuscript, which has now been published in book form.
33. For a history of the rise of a Western and especially North American ecological worldview, see D. Worster, *Nature's Economy: A History of Ecological Ideas 2nd edition* (NY: Cambridge University Press, 1996). This ecological worldview, which sees all things as interdependent and in flux, has influenced some of the rationales behind organic farming. For an exploration of religious environmentalism/s and how this worldview impacts human-nature interactions, see R. Gottlieb, *A Greener Faith: Religious Environmentalism and Our Planet's Future* (NY: Oxford University Press, 2009).
34. Wendell Berry, *The Art of the Common Place: The Agrarian Essays of Wendell Berry*, Norman Wirzba, ed. (Washington: Counterpoint, 2002), 285. See also Berry, *The Unsettling of America: Culture and Agriculture* (NY: Avon Books) 1977.
35. Chidester, David and Linenthal, Edward. "Introduction" in *American Sacred Space* (Indianapolis: Indiana University Press, 1995), 1-42. They also point out that "nature in America is not a 'natural' but a thoroughly cultural production of space" (ibid: 13). The nature found on and in farm lands is thus in part a cultural production and as this paper labors to show, this nature is impacted by cultural and individual understandings of religion--including concepts of a sacred/profane dichotomy.
36. M. Ableman, *Fields of Plenty: A Farmer's Journey in Search of Real Food and the People Who Grow It* (San Francisco, CA: Chronicle Books) 2005, 69.

INSIDE/OUT: THE BODY UNDER ATTACK IN AMERICAN POPULAR CULTURE

Julien R. Fielding

For many decades, Hollywood has terrified its audiences with the suggestion that parasitic entities were poised to take over our bodies and minds. All writers had to do was couple our natural aversion to foreign invaders with our fear of the unknown and *voila* – you get a winning formula that gets regular play in projects that are destined for the big and small screens. During the 1950s, alien invasion films became a staple of the drive-in theater. Many of them begin with a meteor crashing to Earth. Someone, often a scientist, makes first contact and attempts to communicate with the aliens. Sometimes these beings arrive and bring with them a message of peace, but usually they have come for less than benign reasons, typically they want to take over the planet. Because of the way they look, or because their bodies aren't as effective on Earth, the first thing that aliens must do is find a "host," someone whose body they can invade and, eventually, whose mind they can enslave. Although this scenario may seem to be far-fetched, it is actually something that we can witness in nature. Parasites, defined as an organism that lives on or in a host without benefiting said host, can make people go blind, can change how a host eats and digests its food, and can even alter the host's genetic code, making it a better place to live. Some parasites even "control their hosts, becoming in effect their natural brain, and turning them into new creatures. It is as if the host itself is simply a puppet, and the parasite is the hand inside."[1]

This essay will attempt to offer a short survey of parasites in film and on TV. The first section will examine science fictions films in which alien life forms invade human hosts, usually, but not always with the purpose of taking over our planet. (The creature in *The Thing* never states its motive.) Movies in this section, which range from *It Came from Outer Space* (1953) to *The Faculty* (1998), have drawn upon our fear of the unknown and the "other," and, specifically, have exploited their audience's fears of communism and conformity. As one scholar writes: "The individual's fear of being swallowed up by a faceless mass and becoming a drone was a prime concern in 50s sci-fi movies," undeniably sparked by the fear of communism, McCarthyism, and totalitarian political systems.[2] The second section will look at parasites on TV, specifically in the *Star Trek* universe, on *Stargate*

SG-1, on *The X-Files*, on *Buffy the Vampire Slayer*, on *Primeval*, on *Fringe*, on *Sanctuary* and on *Futurama*. The essay will then turn to films in which nature has either "gone wild" or has been given a detrimental nudge by "mad" scientists. Selections range from The *Tingler* (1959) to *The Thaw* (2009). Usually, cinematic parasites are insects, most commonly slugs or worms, and yet, there exist a few films in which the parasitic organism is also a plant. The essay will conclude by looking at two such films, *Creepshow* (1982) and *The Ruins* (2008).

Alien Invasion Films: "Psychic" Parasites

The first alien invasion film to consideration is *It Came from Outer Space* (1953), which is based on a Ray Bradbury story. It begins with a crash landing in the Arizona desert. An amateur stargazer, John Putnam (Richard Carlson), and his girlfriend (Ellen Fields) go to investigate. It doesn't take long before the locals begin acting strangely, usually after they've encountered an unusual kind of mist. The people aren't acting like themselves; they speak in a monotone voice and seem emotionless. As the sheriff remarks: "I couldn't even be sure that you are John Putnam standing beside me." "That's right," John replies, "you couldn't. Wouldn't it be a fine trick if I weren't really John Putnam at all; something from another world come here to give you a lot of false leads." The aliens aren't parasitic per se, because instead of invading the person, taking it as a host, they take the person hostage and then temporarily transform themselves into that person's replica so they can move around freely. When finally confronted, an alien explains that they "have souls" and are "good." All they want is time to repair their spaceships so they can return home. Why have they assumed human form? Because, as the voice continues, "you would be horrified at the sight of us," and indeed when John sees the being's true form, he recoils in terror. John understands that human beings kill what they don't understand, and so urges everyone to give the aliens the time they need. Still, few share his compassion. Matt begins ranting: "How do we know they're not taking over? They could be all around us and I wouldn't know it." In his hysteria he mounts an armed mob and begins shooting at anyone he suspects of being an alien. Even though they came in peace, the aliens decide that they have no choice but to respond violently.

Invaders from Mars[3] is yet another "lost identity" film also released in 1953. In it, a blonde-haired, freckle-faced boy, David (Jimmy Hunt), witnesses a spaceship landing in a nearby sandpit. His father, George (Leif Erickson), an engineer, goes out to investigate. When he hasn't returned by the next morning, his wife, Mary (Hillary Brooke), alerts the police. While they are looking for the missing man, the officers also disappear into the sandpit. Eventually, George returns, but his personality has changed. This formerly kind, benevolent man has become argumentative, dictatorial and violent. Could this have anything to do with the mark that has appeared on the base of his skull? The police eventually return, and they, too, are acting strangely. Mary is the next to disappear into the sandpit, and she is followed by more townspeople, including the chief of police and top military officials. Terrified by what's happening, David tries desperately to find someone to believe him, and eventually, he does – an astronomer, Dr. Stuart Kelston (Arthur Franz), and Dr. Pat Blake from the City Health Department (Helena Carter). Towards the end of the film, David and Dr. Blake are sucked into the sandpit and are abducted by the Martians – giant, green, bug-eyed humanoids – who themselves are controlled and enslaved by a highly "evolved" being that is essentially a silver-colored humanoid, from the shoulder blades up, with tentacles coming out of its shoulders, and an enormous cranium. (It never speaks. Instead it communicates through telepathy.) As for what has happened to all of the townspeople … the Martians seem to be inserting some kind of device into their victim's neck with which they can control the person's mind. No doubt once it is removed, they will return to normal.

Invasion of the Body Snatchers (1956), which is based on Jack Finney's story, is probably the best-known, and most frequently remade example of the "lost identity" film. Not long after Dr. Miles Bennell (Kevin McCarthy) returns home to Santa Mira, Calif., he hears strange reports about people not being themselves. For example, Wilma, the cousin of Bennell's girlfriend, claims that her uncle might be an imposter. "There's something missing … Always when he looked at me there was a special look in his eye – that look's gone … There's no emotion, none. Just the pretense of it." Over the course of the film Bennell will discover that alien pods are giving birth to fingerprint-less and undifferentiated bodies that come alive only when people fall asleep. Once the pod person wakes up, he or she is physically identical to his or her original with the added benefit of being interconnected with all of the other pod people. As one character explains, the humans are "hosts harboring an alien form of life; a cosmic form that, in order to survive, must take over every human being." The film never explains what happened to the original person once he or she is "replicated;" we have to wait for the three subsequent remakes for further exploration of this issue.

Perhaps as a way to fill in the blanks, the 1978 version of *Invasion of the Body Snatchers* begins in outer space on a barren planet inhabited by jellyfish-like creatures. With nothing left to sustain them, they abandon the planet and float into space. Carried by solar winds, they land on Earth and, in the form of palm-sized, blooming pods, begin to spread. Elizabeth Driscoll (Brooke Adams), a scientist at the San Francisco Health Department, is one of the first to find a pod. After plucking it off of a bush, she takes it home and tries to figure out what it is. (Her best estimate is that it's two species that have cross-pollinated to create a third.) That night, she and her boyfriend go to sleep; by his side of the bed is the pod flower in a glass of water. In the morning, she wakes up to find him uncharacteristically dressed in a suit.[4] As in the original film, the average person is seemingly helpless against the alien menace, this time referred to as "space flowers" that initially grow parasitically on other plants.

By the aforementioned definition are these pods "parasites?" To answer this, we have to consider what the plants are doing to the human race. Once a person goes to sleep, a nearby pod attaches filaments to him or her. Next a giant flower blooms and not much later, the pod spits out a slimy human fetus that is fashioned into a perfect copy of the original person. This original eventually disintegrates into a fibrous matter that is seen throughout the film being dumped into garbage trucks. One might argue that these "new" people are nothing more than clones, except for the fact that their memories are intact. Is this hybridization taking place? Is it like a caterpillar becoming a butterfly? As one character explains, one shouldn't be "trapped by old concepts … we're evolving into a new life form." The creatures came from a dying world, and in order to survive on Earth, they are adapting.[5] How are the pods parasitic? They assume our forms and take over our minds and consciousness so that they can manipulate others and survive on a new planet.[6]

The Thing From Another World (1951) is another invasion film from the 1950s, but unlike the previous examples, it didn't feature people being taken over or controlled by an alien entity. This parasitic bent would have to wait until the 1982 John Carpenter remake, *The Thing*. Set in Antarctica, *The*

Thing opens with the crash landing of a spaceship. This scene quickly segues into a sequence during which a malamute is running through the snow. It is being chased and fired upon by a Norwegian in a helicopter. The dog reaches the American National Science Institute Station 4 and is rescued by the men living inside. Unfortunately, as the doomed men will discover, the animal isn't what it appears to be. It is, as Dr. Blair (Wilford Brimley) explains, an "organism that imitates other life forms, and it mutates them perfectly."

The back-story of *The Thing* is identical to the one found in *The Thing from Another World*: Thousands of years ago an alien crashed on our planet and was immediately frozen under the ice. A team of scientists finds the creature's ship and uses thermite charges to get to it. After discovering an alien life form,[7] they take its ice-encased body back to their station. It gets loose and immediately threatens everyone's lives. In the original film, the alien is a Frankenstein monster-like creature that is, in fact, a vegetable. In the remake, it's something less definable. In John W. Campbell Jr.'s novella *Who Goes There*?, which is the source material for both films, we read that this creature "digested" the dog, and as it did so, it "studied every cell of his tissue, and shaped its own cells to imitate them exactly" … "the creature … grew out of the dog."[8] MacReady (Kurt Russell) explains that "this thing doesn't want to show itself. It wants to hide inside an imitation." Appropriately, the film's tagline is "man is the warmest place to hide." Much like the pods in *Invasion of the Body Snatchers*, this thing replicates whatever it encounters as a means of survival. It is a dog but not a dog; it is a man but not a man. At the end of the film, the creature reveals, layer after layer, many of the life forms it has encountered: plants, animals and everything in between.

Finally, there is *The Brain from Planet Arous* (1958), which, like *Invaders from Mars*, features a giant brain that essentially acts as a psychic parasite. It begins with Steve March (John Agar), a nuclear physicist, noticing unusual radiation levels coming from Mystery Mountain, a veritable no-man's land where temperatures skyrocket into the triple digits. Despite protestations from his girlfriend Sally (Joyce Meadows), he and his colleague, Dan (Robert Fuller), drive into the desert to investigate. They eventually come upon a newly created cave, and while inside encounter Gor, an alien brain poised to conquer the world. The meeting proves fatal for Dan. Steve lives but his fate isn't much better. Gor takes over his "thoughts, body and emotions," and moves him around like a puppet. Their relationship could be described as parasitic, because Gor somehow gets inside of Steve, with the result being excruciating pain. Later in the film, another alien brain arrives with the purpose of taking Gor back home. It uses Sally's dog, George, as its host.

Alien Invasion Films: Physical Parasites

So far, we've looked at films in which the aliens could be described as "psychic" parasites. In *The Brain Eaters* (1958), the "aliens" are actual parasites. Although Robert A. Heinlein doesn't get a credit and neither does his novel *The Puppet Masters,* this film is essentially an "unauthorized" retelling of that sci-fi classic. The story begins in Riverdale, Ill., where a 50-foot-tall, cone-shaped vessel is discovered and its appearance leads to in a bunch of dead animals and some brutal murders. Several prominent citizens also have go missing. One of them, the Mayor, later returns but with a slug-like creature attached to the back of his neck. During the Mayor's autopsy – he was shot to death – a local scientist, Dr. Paul Kettering (Ed Nelson), discovers that the creature that was attached to the Mayor has two piercing instruments that allow it to attach to its host. Not only that, but it produces an acid when injured, and, if cut in half, becomes two independent beings. Most importantly, once the alien attaches itself to its host, it takes over that person's mind; he or she becomes robot-like. In the end, we learn that these creatures are 200 million years old and date to the "carboniferous age," which is the "age of very large insects." Somehow these creatures survived inside of the Earth, and have a society that is free from strife and turmoil. (Probably because they are all of one mind.)[9] Kettering is unimpressed with promises of social harmony, and he puts an electrical wire across the top of the cone and electrocutes everything inside.

Thirty-six years after *The Brain Eaters* was made, *The Puppet Masters* (1994) appeared in cinemas across the United States. It too begins with the UFO landing, this time in Iowa. A secret government paranormal agency, led by the no-nonsense and dictatorial Andrew Nivens (Donald Sutherland), comes to town to investigate. As the film develops, we learn that manta ray-like slugs, which are comprised of 60 percent brain, attach themselves to a host's spinal column and tap directly into the person's brain. This allows them to control the person like a puppet. Everyone who has an alien attached to him or her can communicate with the others; everyone is essentially many parts of one entity. Interestingly, the creatures come out of eggs, and they can move around, usually when threatened, by a kind of proboscis. Again, these parasitic aliens have come to Earth to rebuild a new home.

Brain Damage (1988), which is written and directed by Frank Henenlotter, has a lot in common with *The Puppet Masters*. It begins with an elderly couple bringing brains to Aylmer, their bathtub dwelling, tentacled, very talkative worm-like "pet." But, sadly for them, Aylmer has vacated their apartment and has found himself a new host, Brian (Rick Hearst). All Aylmer wants are human brains, but because he isn't ambulatory himself, he needs a host that can bring him within reach of his prey. So how does he convince these people to let him ride around on them? Inside of his toothy mouth are two "hooks" that he inserts into the neck of his host. Through these "hooks," he injects a highly addictive hallucinogenic blue substance that results in the host experiencing bright lights and a feeling of ecstasy. As Aylmer says, "I'm all you'll ever need. I'll show you the light if you take me for a walk. I will do all of your thinking for you." At one point, Brian attempts to get the drug out of his system, but it is hopeless. He is hooked. Aylmer, it is explained, is an ancient being that has been the object of desire since the Fourth Crusades. Where he came from before that, we are never told. On the one hand the low budget *Brain Damage* satisfies its audience's desire for sex, violence and gore, but on the other hand, it makes a "statement" about the addictive nature of drugs and shows us the atrocities one might be willing to commit while under their influence.

Also from this era is the zombie/serial killer/alien parasites hybrid film *Night of the Creeps* (1986). Because it's essentially an homage to the alien invasion film, it begins in 1959 and is filmed, at least initially, in black and white.[10] While a college couple is at a "make-out" spot, they witness the arrival of an alien experiment that has been jettisoned from outer space. Johnny, a fraternity member, goes to investigate and becomes infected with a slug-like creature. An axe-wielding psychopath murders his girlfriend, and he, for some reason, ends up in cryogenic stasis. Flash forward to 1986: two hopeless nerds, Chris (Jason Lively) and his best friend J.C. (Steve Marshall), are on campus, during Pledge Week, when Chris spots a beautiful sorority girl. Believing that the only way he can date her is to join a fraternity house, he and his friend try to pledge. Although there is no chance that these guys will be welcomed into the fraternity, they are told that they will be if they steal a cadaver and dump it on the front steps of the rival fraternity. Chris and J.C. seeking to fulfill this task find a body – unfortunately the frozen, infected frat guy from the beginning of the film– and thaw him out. Thus, unwittingly they release the alien parasite into the community, creating a town filled with zombie-like creatures. In the end, Chris and Det. Cameron (Tom Atkins) discover that fire is effective in killing the creatures and retake the town.

Written by Kevin Williamson, *The Faculty* (1998) is also a hybrid film, paying homage to such sci-fi classics as *The Puppet Masters, The Thing,* and *Invasion of the Body Snatchers* on the one hand; on the other it deals the challenges of average high school student, like bullying, identity, sexuality, and even drug use. A parasitic alien life form lands in small town Ohio with world domination on its mind. The first victim is the bad tempered and aggressive coach (Robert Patrick), who, in turn, "infects" another member of the faculty, and so on, until the whole student body is taken over. Six students from different cliques come together to fight the alien menace before it spreads. The alien itself is a toothy sea-dwelling amphibian that can replicate itself fairly rapidly. In fact, it uses its "tentacles" to go into its victim's ears. Once inside the host's mind, it takes over and connects the person to everyone else. As one character explains, the organism can survive independently, but it nevertheless needs access to a host. Because the organism is aquatic, the host must consume vast amounts of water to keep hydrated. And thus, the students find a way to kill it—by using a diuretic to dry it out. One of the characters, Stokely (Clea DuVall), is knowledgeable of sci-fi literature, particularly *The Puppet Masters*, so she becomes the group's "expert" on how to dispatch the growing menace. She suggests that maybe if they kill the "leader," the subsequent victims will be freed.[11]

Underrated, *The Hidden* (1987) offers a different perspective on the world dominating, parasitic alien. In this film, once a murderous slug-like creature takes over a host, it acts like the *id* gone wild. It takes anything it wants and does whatever it wants. Its preferred activities include stealing Ferraris and driving them at high speeds, having a gun and using it on anyone who gets in its way, and listening to loud music. The only "person" who can stop this ethically challenged entity is another alien hiding in the body of Lloyd Gallagher (Kyle MacLachlan), an FBI agent.

Dreamcatcher (2003), which is adapted from a Stephen King novel,[12] centers on four childhood friends who are at a cabin in the woods, engaging in some male bonding. Two of them are out hunting when they encounter an older man lost in the snow. Because he's been out wandering in the woods, he has consumed some berries, which, he says, have given him indigestion. As our protagonists will discover, the berries have nothing to do with his condition. In reality, a lamprey eel-like creature with hundreds of teeth has invaded his body. It later bursts violently out of his backside while he's sitting on the toilet, leaving behind a mass of blood and the man dead. Another one of these parasitic "shit weasels" is inside of the man's wife; she is found by the other two men who were on a beer run. The ultimate plan of the alien species, whose code name is "Ripley,"[13] is to deposit its wormy offspring into the reservoir that provides Boston with its drinking water, thereby infecting that large population.

Finally, written and directed by James Gunn, *Slither* (2006) begins with a meteorite crashing to the Earth. A local businessman, Grant (Michael Rooker), who is out in the woods, goes to investigate. He finds a maggot-looking creature that shoots a barb into his chest. This barb works its way to the base of his brain, and over time it changes him into a grotesque, carnivorous squid-like creature. Wanting to propagate, Grant goes over to the home of his mistress. He essentially rapes her with tentacles that come out of chest and pump fluid into her stomach. Once she's impregnated, he stashes her in a barn, where her egg-like body incubates the alien spawn. Eventually, her body bursts open, releasing thousands of red, slimy slug-like aliens into the world. They immediately head for town, and whenever they encounter someone, dive straight into the person's mouth. As we find in so many other alien invasion films, once someone is taken over by one of these alien slugs, they become "one" with the other victims.[14]

Parasites on TV

Thus far, our discussion of world-conquering parasitic aliens has been confined to the big screen, but this doesn't mean that they can't be found on television. In 1967, the crew of the U.S.S. Enterprise confronted parasitic creatures that had spread mass insanity to a series of Federation colonies. The *Star Trek* episode, titled "Operation … Annihilate!," begins with the Enterprise discovering a lone Denevan pilot who is steering his ship into the sun. They try to make contact, but it's to no avail; he kills himself. Wanting to find answers, Captain Kirk (William Shatner), Spock (Leonard Nimoy), Scotty (James Doohan) and Dr. McCoy (DeForest Kelley) beam down to the planet Deneva, only to find themselves confronted by a group of hostile men. Screams of "They are here! They are here" lead them to Kirk's dead brother, his hysterical sister-in-law, and his unconscious nephew. The away team beams the survivors to the ship and attempts to determine the cause of the problem. They discover that beige amoeba-shaped creatures, which can fly and are impervious to being shot by a tazer, arrived on a cargo ship more than six months ago. Once they attack a host – Spock becomes one of their victims - the creatures leave behind tentacles that entwine and grow all around his or her nervous system. As Dr. McCoy explains, "it leaves a stinger, much like a bee or a wasp, leaving one of [the tentacles] in the victim's body. It takes over the victim very rapidly. And the entwining is far, far too involved for conventional surgery to remove." These parasitic creatures control their hosts by causing them excruciating pain, which is why the hosts go insane. Spock uses the power of his mind to overcome the pain – he is Vulcan, after all – and returns to duty. Beaming back to the planet, he captures one of the parasites and returns with it to the Enterprise. Spock examines the creature and determines that it is "a one-cell creature, resembling … a huge, single brain cell." In essence, it is one part of a much larger organism, guided by the whole and controlled by the whole. After further exhausting examination, they determine that the only way to kill the creature without harming the host is to use ultraviolet light.

While still in the original *Star Trek* universe, it's worth mentioning that the villain of *Star Trek II: The Wrath of Khan*, Khan Noonien Singh (Ricardo Montalban), uses a parasite to gain the cooperation of Captain Terrell (Paul Winfield) and Commander Chekhov (Walter Koenig). These creatures, Ceti eels, look a bit like ant lions, and "incubate their larvae within the plates of their jointed carapace. Upon emergence, the eel larvae can enter the ear of a larger animal, where it wraps around the cerebral cortex." The result is extreme pain for the host, which renders him or her "susceptible to outside suggestion."[15]

Parasitic worms return to the *Star Trek* universe, this time in the franchise's next incarnation: *Star Trek: The Next Generation* (1987-94). In the season one episode titled "Conspiracy," Data (Brent Spiner) reviews the logs of Starfleet Command and discovers that an unusual amount of personnel shifting has taken place, especially at the highest levels. He theorizes that a "hostile force" is trying to gain control of "important sectors of Federation territory."[16] Commander Riker (Jonathan Frakes) encounters this hostility first hand, when an admiral, who wants to show him an alien life form, attacks him. Subdued and taken to sickbay, the admiral undergoes a physical examination. They discover a gill-like appendage sticking out of the base of his neck. As the episode continues, we learn that the parasitic aliens control the host's mind, and can infect a host by crawling in and out of his or her mouth. In the end, a "mother parasite" emerges, and when she is killed, so are the smaller parasites.

In the season four episode titled "Identity Crisis," the crew of the Enterprise is investigating the disappearance of 49 colonists from the Tarchannen III outpost. An away team, consisting of Lt. Commander Geordi La Forge (LeVar Burton) and Susanna Leijten, beam down to the surface to investigate. It isn't long before the latter officer takes ill, and she is beamed to sickbay. Upon investigation, Dr. Beverly Crusher (Gates McFadden) discovers some abnormalities with the officer's blood chemistry, and as time passes, the sick woman develops a strange blue growth pattern on the back of her neck, her fingers begin to fuse, and she develops sensitivity to light; the patterns eventually spread over her body. Not much later and Geordi, too, begins exhibiting similar symptoms. Dr. Crusher eventually determines that an alien parasite moves into the host's thymus, and while there, uses the host's immune system to transform that person into a new creature.

Not all wormy aliens are villainous in *Star Trek*. During season four[17] of *Star Trek: The Next Generation*, we are introduced to a worm-shaped organism that inhabits the abdomen of its host. Even though these life forms are sometimes referred to as parasites, in actuality they live in a symbiotic relationship with their hosts. When its host dies, the wormy symbiont must be transferred to another host within 48 hours. Once joined, the host gains the memories and personality traits of the symbiont's previous hosts. One of the main the main characters on *Star Trek: Deep Space Nine* (1993-99), Lt. Commander Jadzia Dax (Terry Farrell), is one of these joined beings, known as Trill.

In this series the mythology of Dax's species is developed, especially in the following episodes: In season two, "Playing God" and "Invasive Procedures;" in season three, "Equilibrium;" and in season four, "Rejoined." We learn that because there are more people who want to be hosts than there are available symbionts, a commission has been set up to determine who would and who would not be a suitable candidate. There seems to be one drawback to joining with a symbiont; once the process is complete, there is no way to reverse the process. "The host and symbiont are dependent on each other after 93 hours. If the symbiont is removed, the host will die, even if (he or she) were otherwise healthy."[18]

By contrast, on the TV series *Stargate SG-1* (1997-2007) we find a parasitic, sometimes referred to as symbiotic, alien race called the Goa'uld (meaning God) that "look like snakes but normally insinuate themselves into the brains of large animals, which they can control … Most …. pose as gods to control slave armies and are considered evil, egocentric and megalomaniacs by those who do not worship them." Anyone who becomes host to the Goa'uld suffers miserably, and long-term control can drive him or her insane. If someone tries to remove the parasite, the creature releases a deadly toxin into the host's bloodstream.[19] The Goa'uld is the primary villain in the *Stargate* series, and the alien frequently attempts to turn various characters into hosts for the wormy symbionts.

Early in season one of *The X-Files* (1993-2002), in the episode "Ice," a member of an Arctic ice core drilling team sends a transmission: "We are not who we are. It goes no further than this," and then is not heard from again. Agents Mulder (David Duchovny) and Scully (Gillian Anderson), who are joined by a pilot and three researchers, fly to the research station to investigate. Upon arrival, they discover a number of dead bodies and one living, but very aggressive, border collie. The dog bites the pilot. While tranquilizing the animal, they notice dark nodules under its armpit, much like the kind associated with the Black Death. While autopsying the dead men, Scully finds ammonia in their blood and a single-celled creature. When the agents suggest quarantine, the pilot becomes violent. As they restrain him, they see, under the skin at the base of his neck, a wormy creature. By the end of the episode, Scully has determined that this extraterrestrial worm – it was buried in the ice – inhabits the host's hypothalamus, a region associated with aggression. She also discovers that the only way to "stop" it is for a host to have two worms inside of him or her. Because of the highly territorial nature of the organism, the worms essentially fight to the death.[20]

"The Host," a season two episode of *The X-Files*, begins on a Russian ship off of the Atlantic coast. The sewer lines are blocked, so the captain orders one of the shipmates to fix the problem. After some grumbling, he begins working, only to fall into the tank, which has a "high bacterial environment." His body is flushed out to the ocean. Eventually, his corpse ends up on an autopsy table, with Agent Scully trying to determine what happened to him. During the autopsy, she discovers a fluke worm attached to the bile duct of his liver. She also determines that the parasite was feeding off of that organ. In one of the next scenes, we return to the sewers, where a sanitation worker is repairing a torn drain screen. He falls in and is pulled under by something. When he emerges from the water, he has a giant sucker wound on his back. Before he dies, he coughs up blood and another fluke worm. By the end of the episode, the duo link the attacks to a half primate, half fluke worm. How is such a creature even possible? Radiation. The agents discover that the Russian ship transported waste from Chernobyl.

In "Bad Eggs," an episode that ran during season 2 of *Buffy the Vampire Slayer*, the titular character (Sarah Michelle Geller) finds herself in a scenario lifted straight from Heinlein's *The Puppet Masters* and *Invasion of the Body Snatchers*. To teach his students about the responsibilities of parenting, Buffy's sex education teacher distributes eggs to every student. But instead of yokes and albumin, these eggs contain entities that sport tentacles and take over the minds of their hosts. While the students sleep, the egg-dwelling life forms attach themselves to their hosts' spinal column. Soon everyone in the high school is infected and doing the creatures' bidding save Buffy and her friend Xander (Nicholas Brendon), who discover that these eggs originate with a giant, one-eyed, tentacled "alien" living in the basement of their school building. Once the "mother" is killed, its babies also die.

At least two episodes of the British TV show *Primeval* have featured worm-like creatures; in one of the episodes it is a parasite.[21] During season one, the characters encounter parasite-infected dodo birds that have come through a space/time anomaly. These creatures live in the host's bloodstream before migrating into its nervous system. While there, it makes the host violent. The intention is that before the host dies – none of the birds live very long – this violent behavior will allow the parasite to "infect" a new host. Before dying, one of the dodos bites a human. Almost immediately, he develops an aversion to light and becomes openly hostile.

Created by J.J. Abrams, Alex Kurtzman and Roberto Orci, *Fringe* centers on an FBI agent, Olivia Dunham (Anna Torv), who must work with a brilliant, but previously institutionalized scientist, Dr. Walter Bishop (John Noble), and his son, Peter (Joshua Jackson), to investigate unexplained phenomena. In the season 2 episode titled "Snakehead," a sick Chinese man is looking for a shop in Chinatown. After he finds it, he begins complaining that his stomach hurts. The source of his pain is a giant squid-like worm, which "bursts" out of his mouth and kills him. The man, we learn, had passage on a cargo ship arriving from Hong Kong. Of the people on board this ship, 27 are dead; all are carrying the same parasite inside of their bodies. Upon first glance, Walter claims that the parasites are like those found in livestock, but he's never seen them so large. (One weighs nearly 8 pounds.) During the FBI's investigation, it is discovered that the trafficked Chinese immigrants are being given capsules, supposedly for seasickness, but which actually contain the parasitic larvae. During the journey from Hong Kong to Boston, the vectors develop and grow to enormous size. Once the boat docks, these intestinal hookworms are then "harvested" from the unfortunate humans so as to produce medicine that is eventually sold to rich, white Americans. Walter surmises that these creatures are actually a new species of hookworm, prized for their lymph gland's much sought after immune boosting enzyme, bioengineered and designed to grow especially large in humans.

During season one of the SyFy series *Sanctuary*, and in the episode "Requiem," Dr. Helen Magnus (Amanda Tapping) gets a distress call from the Bermuda Triangle. When she arrives in her submarine, the Nautilus, she discovers that the mermaids, a peaceful race, have been slaughtered. To discover what happened, she brings one of their bodies aboard. During an autopsy, she is infected with a microscopic parasite. Once inside of her, it heads straight to her brainstem and into her limbic system, where it feeds off of her feelings of fear and anger. The longer the parasite is inside of her, the more symptoms she develops, including headaches and erratic and violent behavior. To save the doctor and himself, Dr. Will Zimmerman (Robin Dunne), raises the CO2 levels in one part of the ship, thereby killing her. Once she's "dead," the creature leaves her body through her oozing blood and forms a worm-like shape that slides across the floor. Will freezes it with liquid nitrogen and revives Magnus.

Most of the time, when a character is infected by a parasite in science fiction or horror, the result isn't good. That's not the case in the third season episode of *Futurama* (1999-present) titled "Parasites Lost." Fry buys an egg salad sandwich in the men's lavatory and, despite protestations from his crewmembers, eats it. Almost immediately, his personality changes, as parasites set up a city in his body. But instead of harming him, the parasites actually help him by doubling his thinking power. They also give him a level of sophistication that he had previously lacked. Initially, Fry has no intention of getting rid of his wormy guests, but as the episode continues, he changes his mind. With the parasites gone, he returns to "normal."

Mad scientists

Not all parasitic creatures on celluloid, or on TV for that matter, are alien. Some are the products of mad scientists, and some are simply instances of nature gone mad. *The Tingler* (1959) belongs in the "mad scientist" camp. In it, a pathologist (Vincent Price) who is conducting experimental research discovers that in every human, along the spinal column, there exists an extremely powerful, very real creature that gets its energy and strength from fear. Screaming is the only way to dissolve and/or kill it. Believing that the best way to know the creature is to experience it himself, he drops acid. Unfortunately, he undermines his own efforts when he screams. His "luck" changes when a deaf-mute woman dies of fright, and from her spine, the doctor extracts the parasitic creature, which once it gets loose from its cage, causes all kinds of havoc. Since there's no way to kill it, the doctor decides to bury it with the dead woman in hopes that "when it goes back where it came from, it'll also go back to a thing … infinitely small. Even die."

David Cronenberg's *Shivers*[22] (1975) presents audiences with another "parasite going out of control because of a scientist" scenario. In the film, Dr. Emil Hobbes (Fred Doederlin) claims that he's been breeding a type of parasite that could take over the function of a person's failing organs, when, in fact, he's created a wormy creature that can remove sexual inhibitions. When he realizes that it's simply too dangerous to let live, he kills himself and his test subject, a promiscuous teen. But it's too late. The sexually transmitted parasite is already working its way through the residents of Starliner Towers, creating in them uncontrollable passion and a need to copulate. According to Cronenberg, the idea for *Shivers* sprang from a dream he had about a "spider that emerged from a woman's mouth … while she slept … Afterwards on reflection, I thought 'My God, that image is really giving a physical presence to the idea that things go on within us which are strange and disturbing.' Also it seemed the spider in some way gave her life when she was awake."[23] He further explained that the parasites themselves came out of his "childhood fascination with the microscopic, and with insects."[24]

Directed by Ridley Scott, *Alien* (1979) begins with the seven-person crew of the Nostromo, a commercial towing ship, being awoken out of their cryogenic sleep chambers. The ship's computer, "Mother," has intercepted a transmission from a nearby desolate planet and wants the crew to investigate. Three of them go to the planet's surface, where they find a spacecraft with a fossilized alien at the helm. Its bones are bent outwards, as if it has exploded from within. Kane (John Hurt) finds a hole in the ship, and, on a cable, descends into a cave-like area, where he encounters hundreds of large leathery eggs resting under a thin mist. He slips, bringing him closer to the eggs. As he reaches out to one of them, it begins pulsating. Not long after that, he sees movement inside. Without warning, the top opens and an "organic" life form jumps out and attaches itself to his visor. His crewmates take Kane's body to the ship's infirmary, and the part of his visor that hasn't been burned through, is removed. On the operating table, we see that a "face hugger" completely covers Kane's face; its whip-like tail coils around his throat. Ash (Ian Holm), the science officer, attempts to remove the creature, but realizes that if he does, the man's face will also come off. Trying to determine with what they are dealing, they scan Kane, and see that the parasitic creature has a tube down his throat; it is feeding him oxygen. Dallas (Tom Skerritt), the ship's captain, remarks that not only has the creature paralyzed Kane but it has also put him into a coma. Determining that they must remove the alien, they make an incision below the alien's knuckle; molecular acid comes out and burns through several layers of the ship's hull. Upon further exploration, Ash determines that the outer layer of the creature's skin is protein polysaccharides, and that it sheds its cells and replaces them with polarized silicone, "which gives it a prolonged resistance to adverse environmental conditions." This makes him a particularly formidable enemy. Not much later, Ash, Dallas and Ripley (Sigourney Weaver) go back to check on Kane, and the face hugger is dead and has detached from the man's face. Kane comes out of his coma, a bit disoriented but hungry, and everyone goes down to the canteen for a final meal before going back into cryo-freeze. Kane eats a few bites before he begins choking and then convulsing. As he's violently twisting the table, his chest bursts open and a toothy, worm-shaped alien emerges. It slides across the table before disappearing; Kane is dead. During the film, the alien continues to grow in size, and kills almost everyone with whom it comes into contact. These parasites aren't out to conquer anything, per se; they are simply doing whatever it takes to survive. They are opportunists. The real "villain" in *Alien* is the company that sent these people to their deaths. As the android Ash (Ian Holm) explains to Ripley, the company's primary mission was to bring back the "perfect organism" – "a living being not clouded by conscience, remorse or delusions of morality" - the crew was expendable.

In *Parasite*[25] (1982), which was touted as the "first futuristic monster movie in 3-D,"[26] a scientist (Robert Glaudini) is developing a parasite for the government when, he himself is "infected." With the creature living in his stomach, and causing him excruciating pain, he ventures to a small town to continue his research. Things go wrong, when he meets a gang of punks, who ransack his motel room and take the other parasite, which is being stored in a metal canister. It doesn't take long before the toothy worm gets loose and is on the attack. Very little is said about these parasites except that once they reproduce, they will "release spores" that will attach to more victims. After some experimentation, the scientist learns that he can kill them by using high frequency noises.

Larva (2005), made by the Sci-Fi Channel,[27] is another cautionary tale with an added "ripped from the headlines" feel about it. The small town of Host, Missouri, is essentially owned and run by Fletcher Odermatt (David Selby), the owner of Host Tender Meats. His company provides free feed for the local ranchers and then buys back their livestock. Hoping to increase the growth rate of the cows and to make their meat "tastier," the company's scientists have been developing a new kind of genetically modified feed. But there are problems with it. The new veterinarian in town, Eli (Vincent Ventresca), gets a call from a local rancher about his sick cows. After examining the man's animals, he discovers a parasite unlike he's ever seen. These creatures gestate within a host and then burst out, looking for more blood. The film reveals that the modified feed changes the genetic structure of the parasite. Not only does it make them grow at extraordinary rates but it also causes them to develop an exoskeleton. By the end of the film, these larval parasites resemble giant bat monsters that can leap, fly, and kill humans. No doubt, *Larva* was written as a response to the mad cow disease scare.

Isolation (2005), an Irish film, has a lot in common with *Larva* as it too centers on the dangers of big business interfering with the meat industry. In this situation, Bovine Genetics Technology has paid Dan (John Lynch), a struggling farmer, for the privilege of experimenting on his cow. The company is trying to increase bovine fertility and accelerate beef production. Initially, everything seems to be going well, until the animal is about to give birth. Orla (Essie Davis), a veterinarian assisting on the project, is "bitten" by the calf when she puts her hand inside of the mother. Something isn't right. During that night, the cow goes into labor but can't deliver the calf on her own. Dan makes many attempts but it's no use. He has reluctantly allowed a couple of drifters – Jamie (Sean Harris) and Mary (Ruth Negga) - to stay outside of his desolate farm, and he goes to them for assistance. Once the calf is born, it bites Dan. Orla returns, and kills the calf. During the autopsy, she makes some disturbing discoveries. For instance, the calf is born pregnant with six mutant offspring; their skeletons are on the outside. Orla leaves, but before she does, she tells him to stay away from the creatures. A bit later, Dan finds the mother cow disemboweled – one of the calf's mutant fetuses, which moves around like a worm, was born alive and has been busy infecting the other cows. He heeds Mary's advice and burns the creature. The scientist in charge of the project (Marcel Iures) returns to the farm, and conducts his own autopsy. By the time that he issues a "quarantine," Dan, who was infected through the bite, has already begun exhibiting symptoms, including fever and vomiting. Jamie, too, is infected; Orla is dead. While autopsying her body, the scientist discovers a giant hole in Orla's stomach. The parasite has consumed most of her intestines; it is looking for a new host. As if the situation couldn't get any worse, it does. Looking at a piece left behind in Orla's stomach, the scientist discovers that the parasite now has human DNA inside of it, and it's growing at an alarming rate. "The creature can affect humans … it could infect everyone. People wouldn't know they were infected, but their offspring, mutated and deformed … It could wipe out an entire generation."

Nature Gone Mad

In several films, no one can really be blamed for parasitic activity. It's just horrible consequences of nature out of control. *The Stuff* (1985) is a horror film with a strong message about the consumerism.[28] At Fletcher Mines, an old miner discovers a marshmallow-like substance bubbling out of the Earth. He tastes it and finds it delicious. What's next? Why to bring it to market, of course. The Stuff sells incredibly well. Not only does it taste good, but it also doesn't "spot" when spilled on something. Better yet, it's low in calories. During The Stuff craze, Jason (Scott Bloom) struggles to resist the pressure of his family who exclusively eats The Stuff. He has seen it move, and shouts, "It's alive," and "It's going to kill you; it's going to kill you all." His parents refuse to listen to him, explaining "we don't get tired now that we're eating properly." Jason continues to fight the good fight, and is joined in his endeavor by David "Mo" Rutherford (Michael Moriarty), an industrialist spy who is trying to figure out nature of The Stuff. By the end of the film, we are still in the dark as to the makeup of the substance. What we do learn is that it is prehistoric, it is highly addictive, and has the unfortunate tendency of killing its "host" by bubbling out of the person's mouth and, sometimes, even exploding his or her head. The question remains: "Are you eating it or is it eating you?"

They Nest (Creepy Crawlers) (2000) begins on a ship in the North Atlantic. A man, whose mouth has been taped shut with duct tape, is thrown over the side; his body later washes up on the coast of Maine. Whatever was inside of him, making him ill, has escaped from his mouth, and is looking for a new human host. At first glance, the offender appears to be a Madagascar cockroach, but after some research, Dr. Ben Cahill (Thomas Calabro), the town's newest resident, discovers that it's actually an "African armadillo insect" that inexplicably has decided that humans make better hosts than their typical small animal fare. When the insect attacks, it bites its prey and paralyzes him or her so that it can lay eggs inside. Because of the prolific nature of these insects, the town is soon run over run by "creepy crawlers."

Directed by Lucky McKee for the *Master's of Horror* series, *Sick Girl* (2006) is a short film about Ida Teeter (Angela Bettis), a lesbian entomologist who, one day, receives an anonymously sent parcel from Brazil. Inside is an unclassified, very aggressive, insect. Even though the top of its enclosure is weighted down with a book, it still gets out. One day, it bites Ida's live-in girlfriend, Misty (Erin Brown), and a transformation begins to take place. As it's explained in the film, when the insect bites someone, it injects him or her with a toxin. That toxin contains a protein that modifies the person's DNA so the parasite can lay eggs within the host abdomen. The insect also hunts small prey – it kills a dog – and, from its proboscis, drinks its host's blood.

In *Splinter* (2008), Seth (Paulo Costanzo) and Polly (Jill Wagner) plan on celebrating their anniversary by going camping. When their tent breaks, they scrap the idea and head for a hotel. On the way there, they see a stranded woman (Rachel Kerbs) and stop to help her. Unfortunately for them, it is a trap, and they are soon taken hostage by the woman and her convict boyfriend (Shea Whigham), who is hiding in the trees. The couple's situation worsens when they hit something in the road, resulting in a flat tire. While changing it, the convict gets a splinter in his finger. Unbeknownst to him, he has just been infected by a parasitic mold that slowly takes over its victims, turning them into zombie-like, spike-producing, contorting, fast moving, blood seeking aberrations. Because the creature punctured the underside of the car, the vehicle overheats, and they stop at a gas station to buy coolant, gas and food. Before their arrival, the gas station attendant himself was attacked by a parasite-carrying animal, and is now himself a creature. To get away from him, Seth, Polly and the convict lock themselves in the gas station. While trying to stay alive, the threesome tries to figure out what they are up against. Seth, who is a Ph.D. candidate in biology, determines that the sole purpose of the heat-seeking organism is to find blood, from which it extracts its nutrients; it uses its host for mobility and food. What makes the creature such a formidable enemy is the fact that even if the "host" is killed; the organism survives in whichever body parts remain. For instance, in one scene the hand of the convict's girlfriend is cut off, and yet, it still moves around, seeking sustenance.

The Thaw (2009), which is written and directed by Mark A. Lewis, begins on Banks Island in the Canadian Arctic, where Dr. Kruipen, an environmentalist (Val Kilmer), and his team are conducting field research. After they tranquilize a polar bear, they discover that it was dining on a half-frozen body of a woolly mammoth. They take the polar bear back to the base station, and are surprised to learn that it has died. Soon the other researchers become ill. Worried, Kruipen radios a message that his daughter and three ecology students should refrain from flying in, but it's too late. Not long after their arrival, the helicopter pilot is bitten by an insect, followed by one of the students. The pilot dismisses his small wound as a flea bite, when in reality, it's something much worse. The woolly mammoth and then the bear were infected with eggs of worm-like insects, and humans are now hosts for these creatures. Since a person rarely survives the parasite, his or her corpse becomes food for the insects that then go in search of a new host. The life cycle of the organism is surprisingly rapid, which only increases the threat to humankind.

Parasitic Plants

Cinematic parasites are usually associated with helminthes[29], which include roundworms, tapeworms, and flukes; arthropods, insects, arachnids or protozoa. But in some very rare instances, they can also be flora[30] as in the case of *Creepshow* (1982) and *The Ruins* (2008). In the former film, the segment in question is titled "The Lonesome Death of Jordy Verrill." Played by horror writer Stephen King, Jordy is a not-so-bright farmer who finds a meteor on his property. Thinking that he can sell it for $50, he reaches into the crater, drains the glowing liquid inside, and puts the object in a bucket. As he's watching TV, he notices a green fungus like substance growing on his fingers. Slowly this alien "grass" spreads over his property and his body, until he's a humanoid shaped grass monster. Rather than succumb to his fate, he blows off his head with a shotgun.

We never learn the origin of the parasitic vines in *The Ruins,* but because of what they can do – mimic cell phones and laughter – we can speculate that they, too, come from outer space. In the film, four friends are vacationing in Mexico. While at the pool, they meet two foreigners – one is Greek, the other German - who ask them if they would like to accompany them to an ancient Mayan ruin. The students accept. While making their way to the site, they meet some hostile, weapon-toting villagers. To get away from them, the five survivors – one of the men is shot – climb up a temple covered with flower-producing vines. No one follows them. Over time, the 20-something-year olds become victims to the flesh-eating vines. Considering the fact that these entities are carnivorous, one might wonder why they are being called "parasitic." It is never explained why they do this, but, in several cases, the vines enter the bodies of their victims, usually through a wound, and while inside, they squirm beneath the skin's surface and even appear to grow. When discussing the film, director Carter Smith typically refers to the vines as parasitic: "At the heart of what's unsettling about this movie is this idea of bodily infection. Something that gets into your skin and thrives under the surface ... We looked at so many documentaries and YouTube clips of under-the-skin parasites and they are disgusting."[31]

Concluding Remarks

This essay is an attempt to highlight some films and TV shows in which one finds parasitic creatures. No doubt, this only scratches the surface of what's available on VHS, DVD or online. For instance, because my objective was to stick to the science fiction and horror genres, no mention was made of parasites occurring in episodes of such realistic medical TV series, such as *House M.D.* or *Off the Map*. Nor was there any discussion of documentaries or shorts. Furthermore, because it was outside of the scope of this essay, no mention was made of parasitic creatures that appear in videogames, and there exist quite a few examples of these as well. Also excluded from this survey were some "foreign" TV series, such as *Cleopatra 2525*, simply because they either aren't well known or aren't readily available; also films in which a parasite appears but doesn't play a significant role. For instance, in the recent horror film, *Seconds Apart* (2011), homicidal twins make a priest believe that he has an African parasite in his leg, and in *Cloverfield* (2007), we learn that the monster itself is afflicted with parasites. Hopefully, this essay will serve as a resource for someone who wants to explore why these creatures have been so ubiquitous in popular entertainment or for someone who wants to expand upon this survey.

Notes

1. Carl Zimmer, *Parasite Rex: Inside the Bizarre World of Nature's Most Dangerous Creatures* (New York: The Free Press, 2000), 82.
2. Anne Billson, *BFI Modern Classics: The Thing* (London: British Film Institute, 1997), 64.
3. This film was remade in 1986 by Tobe Hooper
4. He is sweeping up shards from the broken glass – the plant is gone – and perhaps remnants of his former body.
5. The film presents us with a whole host of philosophical questions. What makes us "us"? Are we our bodies? Are we our consciousness? Are we something else? If we no longer have feelings – as these new forms don't – do we cease to be human? Would we sacrifice love and hate for a utopian world free of violence?
6. Two additional remakes of *Invasion of the Body Snatchers* have been made. One, in 1993 titled *Body Snatchers,* contains a scene during which the female protagonist, while in the bathtub, is covered with tentacles from a pod that has been stored in the space above the ceiling. Once the tentacles are taken off of the body and removed from its orifices, the link between the pod and person is broken. In the latest version, *The Invasion* (2007), a space shuttle crashes to Earth releasing an alien virus that infects anyone who touches it. One of the first to be infected is a high level official with the Centers for Disease Control who spearheads a "vaccination" program that turns everyone into emotionless citizens. As it's explained in this film, the organism is a spore of unknown origin that attacks human DNA while the host sleeps. When the person wakes up, he looks like us; talks like us, but has no emotion (and doesn't sweat.) The sole objective of these entities is to infect everyone else and take over the world. It should be said that *The Invasion* wasn't breaking any new ground. The idea of a rapidly spreading alien virus or alien is at the center of Cook's *Invasion* (1997), and the independent sci-fi film *Invasion* (2005). One could say that the Shaun Cassidy produced TV series *Invasion* (2005) was inspired by *Invasion of the Body Snatchers*. We will never know, though, as ABC canceled it after one season.
7. In *Who Goes There*?, when the scientists dig up the entity it has a "face ringed with a writing, loathsome nest of worms, blue, mobile worms that crawled where hair should grow." See Billson, 15.
8. Billson, 52–53.
9. One of the most insidious foes in the *Star Trek* universe, particularly in *Star Trek: The Next Generation* and *Star Trek: Voyager,* are the Borg. Dressed in black "leather," these cyborgs arrive in a cube and quickly assimilate everyone with whom they come into contact. As they are known to state, "Resistance is futile." the Borg are essentially "one" entity, known as the Borg Collective, because everyone is linked into the hive mind. When Captain Jean-Luc Picard is himself assimilated in the feature film *Star Trek: First Contact* (1996) he loses his identity and gets a new one, Locutus of Borg. Apparently, these foes were originally conceived of as insects, but because of budgetary constraints, they became humanoid.

10. *Night of the Creeps* is one huge wink to horror fans in the know. Fred Dekker's screenplay contains just about every convention in the horror/sci-fi playbook, including alien parasites, a serial killer, and zombies. His characters, and a few places, are all named after well known horror directors, whose own films, no doubt, inspired Dekker, including George A. Romero (*Night of the Living Dead*), John Carpenter (*Halloween*), David Cronenberg (*The Fly*), John Landis (*Animal House*), Tobe Hooper (*Invaders from Mars*), and Sam Raimi (*The Evil Dead*). Even producer Roger Corman (*Attack of the Giant Leeches* among so many others) gets a nod.
11. While devising her theory, she specifically references *Invasion of the Body Snatchers* and *The Puppet Masters.* The idea about killing the leader to free all of his/her victims can also be found in vampire and werewolf films.
12. In *The Mist* (2007), based on a short story by Stephen King, a secret military operation known as Project Arrowhead produces a strange mist hovering over the lake. In it one finds any manner of carnivorous, over-sized insects and beasts just waiting to attack and devour the local townspeople. One of the creatures, a giant spider, apparently lays its eggs inside of its victims, because one unfortunate person is found wrapped in a web with creatures throbbing beneath his facial skin. After his body falls to the ground, hundreds, if not thousands, of these parasitic spiders burst out of him.
13. The name is obviously a reference to Sigourney Weaver's character in *Alien* (1979).
14. On the DVD commentary track, the director claims that "*The Thing* was the Bible on the set of *Slither*." One of the most easily recognizable winks to the 1982 film is the Mayor's last name, MacReady, which is also the name of Kurt Russell's character.
15. "Ceti eel," last accessed June 18, 2011, http://memory-alpha. org.
16. For a summary see, "Conspiracy," last accessed June 18, 2011, http://memory-alpha.org
17. The episode is titled "The Host."
18. "Trill," last accessed June 18, 2011, http://memory-alpha.org/wiki/Trill.
19. "History of the Goa'uld," last accessed June 18, 2011, http://www.sg-soc.com
20. The show's creator, Chris Carter, claimed that his "biggest inspiration" for the episode was John Carpenter's *The Thing*, however, writers Glen Morgan and James Wong came up with the idea from a report about a Greenland expedition, during which scientists dug into the ice to an "all-time record depth." See Jane Goldman, *The X-Files Book of the Unexplained*, vol. 1 (London: Simon & Schuster, 1995), 94.
21. In season 2, episode 2, giant, prehistoric carnivorous worms appear in a skyscraper and begin consuming many of the people working there. Arriving with these creatures is a sulfurous fog, which the characters decide is allowing these worms to exist. The solution to the worm problem, they surmise, is to vent the building of the sulfurous atmosphere. Once it's gone, the worms should asphyxiate.
22. Alternate titles for the film include *Orgy of the Blood Parasites*, *The Parasite Murders, Frissons*, and *They Came from Within.*

23. Chris Rodley, ed., *Cronenberg on Cronenberg* (London: Faber and Faber, 1993), 43.
24. Ibid, 47. Furthermore, he said that for a while, he had thought about using real leeches for the close-up shots of the parasites, but plans changed when someone accidentally froze the leeches he was keeping alive for that purpose.
25. There is very little parasitism taking place in the 2003 film, also titled *Parasite.* A team hired to use an experimental enzymatic cleaner on a soon-to-be-demolished oil rig fail to read the mixing instructions, and thus end up creating giant worm-like monsters that eat whomever the find.
26. The film is set in 1992.
27. It is now called the SyFy Channel.
28. At the end of the film, Rutherford realizes that the only way to destroy The Stuff is to burn and cook it. It survives, though. The industrialists repackage it as "The Taste," and are about to launch the product, when Rutherford forces them to eat it themselves. Despite every effort being made to keep the substance out of the people's hands and mouths, we see that it is being sold on the black market.
29. Worms or worm-like creatures appear in a variety of sci-fi and horror films; however, they aren't necessarily parasites. In most cases, they are carnivorous. Some examples include *Squirm* (1976), *Dune* (1984), *Slugs* (1987), and *Leeches!* (2003). In *Tremors* (1990) and its four sequels, the "Graboids," giant underground dwelling creatures that have snake-like creatures in their mouth, may resemble the sandworms in *Dune*, but if one looks at the complete creature, it is not a worm. Finally, another kind of parasite that we see in films is the parasitic human. For instance, during season two of *The X-Files,* in the episode titled "Humbug," a man's "evil" parasitic twin detaches itself from its host and goes on a killing rampage. The killer in *Basket Case* (1982) is a vengeance seeking parasitic twin.
30. Killer plants, often from outer space, can be found in a number of films, including *Godzilla vs. Biollante* (1989), *Attack of the Killer Tomatoes* (1978), *The Day of the Triffids* (1962), *The Little Shop of Horrors* (1960 and 1986), *Minority Report* (2002), and *The Happening* (2008). In the *Star Wars* universe, the Sarlacc is a large arthropod whose beak-like mouth has rows of sharp teeth and tentacles; however, it is also plant-like in that it reproduces by spores. See http://starwars.wikia.com/wiki/Sarlacc for more details. In *Star Wars: Return of the Jedi* (1983), Jabba the Hut intends to have Han Solo (Harrison Ford), Luke Skywalker (Mark Hamill) and Chewbacca (Peter Mayhew) thrown into a Sarlacc. Finally, in two episodes of *The X-Files, Darkness Falls,* from season one, and *Field Trip*, from season six we find carnivorous plants.
31. Lawrence Ferber, "Out Director Carter Smith Helms *The Ruins*," *Edge Boston* (April 3, 2008), http://www.edgeboston.com

List of Contributors

Charlotte Baker is Lecturer in French in the Department of European Languages and Cultures at Lancaster University. She completed a PhD in Twentieth Century French and Francophone Literatures at the University of Nottingham in 2007, and her current research focuses on marginalised and stigmatised groups in sub-Saharan Africa. Charlotte is particularly interested in the fictional work of Guinean writer Williams Sassine and recently published her first monograph, *Enduring Negativity: Representations of Albinism in the Fictional Work of Didier Destremau, Patrick Grainville and Williams Sassine* (Oxford: Peter Lang, 2011).

Yvonne Chireau is Professor of Religious Studies at Swarthmore College, where she teaches courses in American religion, African American religion, and comparative religion. She is the author of *Black Magic: African American Religion and the Conjuring Tradition* (2003). Her research interests include the intersections of religion and healing, religion and race, and religion and magic.

Marta Crivos, Maria Rosa Martinez, Carolina Remorini, and **Anahi Sy** are members of the Equipo de Investigación en Etnografía Aplicada (EDINEA) affiliated with the Facultad de Ciencias Naturales y Museo, Universidad Nacional de La Plata as well as the Consejo de Investigaciones Científicas y Técnicas at La Plata, Argentina. Their team develops ethnographic and trans-disciplinary research about subsistence activities that transect the domestic scope in rural populations representative of the high bio and ethnic diversity of Argentina. The team has extensive research experience in ethonography, ethnobiology, and ethnomedicine in native populations from different regions of Argentina.

Julien R. Fielding teaches in the Department of Religious Studies at the University of Nebraska at Omaha. She is the author of *Discovering World Religions at 24 Frames Per Second*, and has contributed several essays to the *Continuum Companion to Religion and Film*. She is also an associate editor for the *Journal of Religion and Film*, and she has her own film-related web site, fieldingonfilm.com.

Brenda Gardenour holds a PhD in medieval history from Boston University and specializes in the history of medieval medicine and hagiography. Her current research examines the use and abuse of Aristotelian discourse in the medieval world, the development of medicine as a language of authority, and the continued influence of both on the deeper structures of modern mentalités, particularly those linked with the horror genre from gothic literature to digital gaming. She is currently Assistant Professor of History at the Saint Louis College of Pharmacy.

Amber Campbell Hibbs is a biocultural anthropologist whose research examines the influence of human interaction with the environment on health and illness. She has studied irrigation use and other farming practices as they relate to disease transmission dynamics in archaeological populations and is currently developing climate education programs for rural populations including agricultural producers in Kansas and Nebraska as part of the Central Great Plains Climate Education Partnership.

Todd LeVasseur received his Ph.D. in religion and nature from the University of Florida, with his dissertation investigating what he calls "religious agrarianism." He is currently a Visiting Instructor in the Department of Religion at the College of Charleston in South Carolina.

Alison More is a researcher for the project 'Religious Orders and Identity Formation in Late Medieval and Early Modern Europe' at Radboud University in Nijmegen. She obtained her PhD from the University of Bristol in 2005 and has held the post of Assistant Professor at the Franciscan Institute of St. Bonaventure University. She specializes in the devotional climate of the high Middle Ages, and has a particular interest in the role of gender in religious devotion. She has co-edited a volume, *Representing Medieval Genders and Sexualities in Europe, 600-1530.* She is working on an edition of Beguine rules for the Brepols series, *Disciplicina Monastica.*

Mark G. Pitner, PhD is an Assistant Professor of Asian Studies at Elmira College. He has written on developments in Ruism (Confucianism), the work and reception history of Yang Xiong (53 BCE–18 CE), the history of natural science in China, and is currently working on a number of projects that explore the role of place in the intellectual history of China.

Misha Tadd is a PhD candidate at Boston University specializing in Early Daoism. He received a Chiang Ching-kuo Foundation Doctoral Fellowship for his work on *Heshang gong zhu*, a little-studied but seminal *Daodejing* commentary. Through this text, his dissertation explores the intersection of body, religion, and politics, and the idea harmony between the individual and society. Currently, he is an adjunct faculty member at Loyola Marymount University.

Amy M. Thomas is interested in the social fabrics and political histories underpinning medical practices, especially the symbolic dimensions of parasitism. "Worms Politic" draws from her undergraduate research at American University, where she received her B.A. in Literature. She is currently working toward a Ph.D. in anthropology at Princeton.

Lavanya Vemsani holds two PhDs in History and Religious Studies, one from the University of Hyderabad and another from McMaster University. She is currently assistant professor of Asian History and Religious Studies in the Department of Social Sciences at Shawnee State University. She is the author of *Hindu and Jain Mythology of Balarama* and a number of articles on the history and religions of India. Her research interests include Hinduism, Jainism, popular religion in India, early modern women and the early history of India, and both classical and modern literatures of India.

Lesley Jo Weaver is a Ph.D. candidate at Emory University, where she studies medical anthropology, religion, and gender in India. Her scholarly interests include chronic diseases and mental health, leprosy and social stigma, and the interactions between gender, religion, and health. Weaver's current research explores the social pathways connecting type 2 diabetes and mental health problems among women in New Delhi.

www.ingramcontent.com/pod-product-compliance
Ingram Content Group UK Ltd.
Pitfield, Milton Keynes, MK11 3LW, UK
UKHW022253091026
15624UKWH00016B/379

9 781433 115479